# Manual of Purpose-Made
# Woodworking Joinery

To Mary Elizabeth, who enriched my life many years ago by adding Goring to her majestic names. Sadly, though, this transaction caused her to forfeit the magnificent maiden name of Wood.

**Books by the same author:**

First-Fixing Carpentry Manual
Manual of First- and Second-Fixing Carpentry

# Manual of Purpose-Made Woodworking Joinery

## Les Goring, ACIOB, FIOC, FTCB, LCGI, MIWSc

*Fellow of the Institute of Carpenters*

*Former Senior Lecturer in Wood Trades at Hastings College of Arts & Technology*

**Drawings by the author**

Routledge
Taylor & Francis Group

LONDON AND NEW YORK

First published 2014 by Routledge

Published 2019 by Routledge
2 Park Square, Milton Park, Abingdon, Oxon OX14 4RN
52 Vanderbilt Avenue, New York, NY 10017

*Routledge is an imprint of the Taylor & Francis Group, an informa business*

British Library Cataloguing in Publication Data
A catalogue record for this book is available from the British Library

Library of Congress Cataloging in Publication Data
Goring, L. J.
Manual of purpose-made woodworking joinery / Les Goring.
pages cm
Includes index.
1. Joinery. 2. Woodwork. I. Title.
TH5663.G67 2013
694'.6--dc23
2013023810


ISBN: 978-0-415-63683-4 (pbk)
ISBN: 978-1-315-85881-4 (ebk)


Typeset in Adobe Caslon and Futura by
Servis Filmsetting Ltd, Stockport, Cheshire

# Contents

# Preface

This book was written as a sequel to my last book entitled *Manual of **First- & Second-Fixing Carpentry***, which was prompted by a perceived lack of trade books with a strong practical bias, using a DIY step-by-step approach – and not because there was any desire to add yet another book to the long list of woodworking books already on the market. Although many of these do their authors credit, the bias is usually from a technical viewpoint only, with wide general coverage; and I wanted my books to be *manuals* that deal with the sequence, techniques and practise of performing the various, unmixed specialisms of a trade.

Such is the aim of this book, then, to present a practical guide through the main items classified as *purpose-made joinery*. This italicized term refers here to items of joinery that are made-to-measure by hand and/or by using portable powered- or fixed-machines in small to medium-sized workshops, as opposed to large, more mechanized workshops or factories with CNC (Computer Numerical Control) and CAD/CAM (Computer-Aided Design and Manufacture) machines. The items covered here include rods and setting out, joinery joints, traditional and modern wooden casements with hinged sashes and box-frame windows with vertical, sliding sashes (still preferred by many owners of period properties), doors and doorframes, stairs and staircases and shelving arrangements, etc.

The book should be of interest to a variety of people, but it was written primarily for craft apprentices (whose diminished numbers are hopefully due to increase in the near future), trainees and building students, established trades-people, seeking to reinforce certain weak or sketchy areas in their knowledge and, as works of reference, the book may also be of value to vocational teachers, lecturers and trade instructors.

Also, the very detailed, step-by-step, graphic treatment of each subject should appeal to the keen DIY enthusiast and the hobbyist.

Finally, with reference to a few reviewers' well-meant comments on the draft copy of the stair chapter in this book – such being comments aimed at the entire proposed book – the final presentation has not been dumbed down to suit beginners, because I believe that they should raise their efforts and pit their wits against what needs to be studied in their subject area.

*Les Goring,*
*Hastings, East Sussex*

# Acknowledgements

In no particular order, the author would like to thank the following people and companies for their co-operation and help in various ways. Gavin Fidler, Joanna Endell-Cooper, Matt Winkworth and Emma Gadsden, Editorial staff at Taylor & Francis, publishers; Axminster Tool Centre; My son, Jonathan, for computer advice, etc; My daughters, Penny and Jenny, for their moral support; Olly Adams, for proof-reading and modelling at the surface-planing machine; My son-out-of-law, Darren Eglington, for computer advice, etc; My brother, Arthur, for his professional involvement over the 'phone; Ironmongery Direct; Mark Gardiner of Croft Glass Ltd, Hastings, for glazing information; Lucy Woodhead and Jane Vincent of Jeld-Wen Windows & Doors; Rachael Tranter of Ventrolla Sash Window Renovation Specialists; Sash Window Conservation Ltd, of Staplehurst, Kent; Mark Kenward, friend and Building Surveyor; Harry Chalk, friend and former lecturer at South East London Technical College; Kevin Hodger and Peter Shaw, former lecturers at Hastings College of Arts & Technology (HCAT); Tim Hedges, former joinery student, now a director of Howard Bros Joinery Ltd, Battle, East Sussex; Peter Oldfield, formerly of HCAT; Andy Cryar, for updated stair-fixing information; John Taylor, formerly of HCAT; Russell and Lesley Griffiths, for kindly allowing me to photograph their staircase that I designed and built in 1995; the following publisher's reviewers of the sample stair chapter of this book: Glenn Whitehead; Thornton Smith; Simon Bilton; P. J. Clancy; Andy Cranham; Colin Fearn; and last, but not least, my own reviewer, Tim Hedges.

# Health and Safety Awareness

Hand tools, power tools and machines are obviously mentioned in this work, but the book does not purposely set out to instruct readers in the safe use of them and where potential hazards lie. And those hazards – like so-called 'grabbing' on narrow band-saw machines, or 'kick-back' on circular-saw machines, etc, lie there cat-like for the ignorant or blasé operator who does not develop a healthy awareness of the potential personal danger involved. However, safe practices are mentioned here where the opportunity allows, but it is essential that readers without any formal training in the use of portable and/or fixed woodworking machines treat them as you would a strange cat. But the difference is to treat them like a strange cat forever. Never get too familiar with them, because (to use an old cliché) familiarity breeds contempt – and you need to respect the power of the machine ad infinitum.

With or without power-tool/woodworking-machine training, educate yourself with the principles of woodcutting machinists' work by reading textbooks on the subject – and *always* read the manufacturer's literature and instructions on portable power tools or fixed machines that you acquire or use.

Finally, all users of power tools and machines should read up on the legislation concerning machinery in the 'Provision and Use of Work Equipment Regulations 1992' (PUWER) and the 'Woodworking Machine Regulations 1974'. The latter deals specifically with woodworking machinery, its safe use and guarding and is, therefore, essential reading.

# 1

# Softwoods and hardwoods
## *Timber defects*

## INTRODUCTION

Woodworkers do not need a degree in wood science or biology, but they do need a certain basic knowledge regarding the converted (sawn) timber from different commercial species of felled trees, its often unruly behaviour and defects and its suitability or not for certain jobs. Such basic botanical knowledge – aimed at briefly here – is an essential foundation upon which to develop an understanding of working with this fascinating and rewarding material.

## GROWTH AND STRUCTURE

Trees are comprised of a complex structure of microscopic cells which take up sap (moisture containing mineral salts) from the soil via the roots, through the sapwood to the branches and the leaves, where it is converted into food and fed back to the new, inner bark (or *bast*) in the cambium layer. The cambium layer conveys the food to the growing parts of the tree via the medullary rays. Another group of cells, which form the bulk of the hardwoods' structure, are the *fibres*. These provide mechanical strength to the structure of the tree. Botanically, the tubular cells of softwood are called *tracheids* and the cells of hardwood are called *vessels* or *pores*.

## CLASSIFICATION

Trees are classified botanically, using two-worded Latin names such as *Araucaria angustifolia* (Parana pine), *Quercus robur* (English oak), etc; and they are also classified commercially into *softwoods* and *hardwoods*. However, these universally accepted trade names must not be taken too literally, as some softwoods are hard and some hardwoods are soft. Parana pine, for example, is botanically a softwood which is quite hard – and Obeche *(Triplochiton scleroxylon)*, a

hardwood, is quite soft. The names really refer to the different growth and structure of the trees and their use in a commercial sense.

## Softwood

Softwood refers to timber from *coniferous trees* – cone-bearing trees with needle-shaped leaves – which are mostly evergreens and are classified as *gymnosperms*. These include the pines and firs, etc. Such species are used extensively for joinery and carpentry work and are marketed as being either *unsorted* or graded according to the straightness-of-grain and the extent of natural defects (described later).

## Redwood and whitewood

Softwood is also described as being either *redwood* or *whitewood*. Redwood refers to good quality softwood such as Scots pine, Douglas fir, red Baltic pine, etc, which has very close, easily discernable annual rings denoting slow, structural growth (a necessary ingredient for strength), a healthy golden-yellow or pinkish-yellow colour and a good weight – and *whitewood* usually refers to a poorer quality of softwood such as a low grade of European spruce *(picea abies* from the *pinaceæ* family), which has very wide (barely discernable) annual rings denoting fast growth and a lack of the required thick-walled, strength-giving cells (tracheids), a pallid, creamy-white colour and an undesirable light weight. It usually has hard glass-like knots – which splinter and break up easily when planed or sawn – and it is not very durable.

## Hardwood

Hardwood refers to timber mostly from *deciduous trees* that have broad leaves which they shed in autumn. These are classified as *angiosperms* and include afrormosia, iroko, English oak, the oaks from other countries, mahoganies, beeches, birches, etc. The first three of these named hardwoods are very strong and durable and are often used in the manufacture of external joinery such as doors, door- and window-sills, door- and window-frames, etc. Of course, for aesthetic reasons, hardwoods from a wide variety of species are also used for internal joinery.

# BOTANIC TERMINOLOGY

## Medullary rays

*Figure 1.1(a)(b)(c)*: This term refers to thin bands of cellular tissue, also called *parenchyma* or *pith rays*, which serve as storage for food which is passed through *ray pits* to the *tracheids* for distribution. These rays radiate from the *pith* in the centre of the tree, to the bark. The rays become very decorative when exposed superficially in certain timbers such as quarter-sawn oak, producing an effect known as *silver grain*.

## Sapwood

*Figure 1.1(a)*: This is the youngest, active growth of a tree, occupying a narrow or wide band, varying from about 12mm width in some trees, up to half the radial area of the tree's trunk, adjacent to the heartwood which occupies the central area. It can often be detected on the surface-edges of slab-sawn boards as a light blue to greyish blue in softwoods – and

in hardwoods, the colour of the sapwood is usually lighter than the heartwood. Because of the sapwood's open grain and its large amount of sap and mineral content, it has a lower durability than heartwood and is therefore less stable. However, it is still used, but because it is more absorbent than heartwood, its position on external joinery components – such as sills – should (ideally) be on the internal part of the sill.

## Heartwood

*Figure 1.1(a)*: This is the mature central portion of a tree, which – because of the effect of the stabilized content of substances such as gum, resin and tannin – is usually darker than the sapwood. Each year a band of sapwood becomes a band of heartwood, as a new band of sapwood is added from the cambium layer around the tree trunk.

## Cambium and annual rings

*Figure 1.1(a)(b)(c)*: Each year a layer of new wood is formed around the outer surface of the tree, varying in thickness from 0.5mm to 9mm for different species. This growth forms under the bark in the cambium layer and – as mentioned above – this cambium layer becomes an annual ring as it is superseded by a new cambium layer.

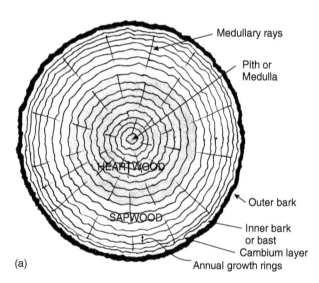

(a)

Figure 1.1 (a) A cross-section through a tree trunk showing a typical area of heartwood and sapwood; annual growth rings; the pith or medulla; the radial medullary rays; the outer bark; the inner bark, bast or phloem; and the cambium layer.

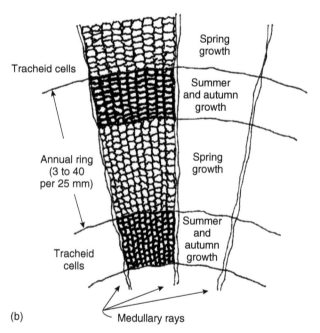

(b)

Figure 1.1 (b) Artistic impression of the magnified cellular structure of softwood.

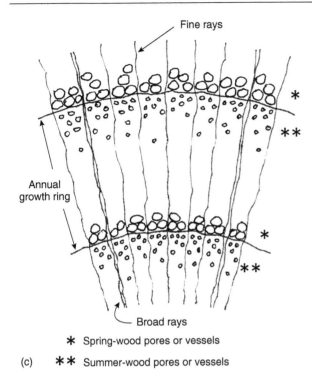

Fine rays

Annual growth ring

Broad rays

✱ Spring-wood pores or vessels

(c)    ✱✱ Summer-wood pores or vessels

Figure 1.1 (c) Artistic impression of the magnified cellular structure of hardwood.

## Spring and summer growth of annual rings

*Figures 1.1(b)(c)*: Most annual rings show themselves distinctly on the end-grain of converted (sawn) timber and this is because the *spring growth* of the rings is lighter in appearance (because the cells are larger) and this contrasts (usually distinctly) with the *summer and autumn growth* of the rings which are darker in appearance (because the cells are smaller and denser).

# SEASONING

*Seasoning* of timber after it is felled and converted (sawn into a variety of sectional sizes), means drying out and reducing the moisture and sap until a certain, necessary percentage remains. This remaining percentage is known as *moisture content (mc)*. If timber is not properly seasoned, it will warp, twist, shrink excessively and be more prone to rot. Commercially, timber is seasoned before it is sold.

The required moisture content varies between 8 and 20% and should be equal to the average humidity of the room or area in which it is to be fixed. 10% mc is suitable for timber in centrally heated buildings; and 10 to 14% in buildings without central heating. External joinery should be 15% mc. Timber for first-fixing carpentry jobs (i.e. roofing and floor joists, etc)

should be 16 to 18% mc; 20% mc being the maximum. Timber with more than 20% mc is liable to be affected by dry-rot and wet-rot decay. 20% mc is generally considered to be the dry-rot safety line.

If the average humidity of a building is less than the moisture content of the timber (as is usually the case after installing central heating), the timber may shrink and split. If the building's humidity is more than the timber's mc (as often happens with condensation problems), the timber may swell and rot. In such cases, the first signs are usually the appearance of a black mould growth on the timber (and other surfaces).

## Methods of seasoning

There are three methods used. These are (1) natural seasoning, (2) artificial seasoning, and (3) a combination of natural and artificial.

## Natural or air seasoning

In common with the other two methods, the branches are removed just after the trees are felled, the trees are then cut into logs and the bark is removed. Eventually, the logs are converted into *baulks* (sized at least 150mm × 150mm) or *planks* (sized at least 300mm × 50mm) and stacked to allow air to circulate around them. The stacks may be out in the open, with rough-boarded roofs over them or in open sheds, having roofs and one or two walls.

Hardwood logs may be converted into planks and stacked one above the other with, say 25mm × 25mm softwood *piling sticks* between them, laid across the planks at maximum 1metre spacing. This is done to create all-round air circulation. Vertically, the piling sticks between the planks must be placed carefully above each other to achieve an equally distributed load; in the drying-out stage, an unequally distributed load may cause distortion of the timber. Also, the ends of the planks are usually painted to prevent end-splitting.

Air seasoning is a slow process and in countries with moderate climates and average weather conditions, it is usually not possible to reduce the moisture content much below 20%. Softwood of 50mm thickness takes approximately 3 to 4 months to reduce to 20% mc. Hardwood of 50mm thickness takes approximately 12 months to dry to the same amount.

## Artificial or kiln seasoning

The converted timber is usually stacked on rail-trucks (with piling sticks between each layer) and placed in steam-heated kilns which force out the moisture and sap from the timber, until the required mc is achieved. Strict control of heat and humidity has to be maintained to prevent defects occurring. Such defects include warping, splitting – and case-hardening (drying too quickly on the outside, causing moisture to be trapped on the inside).

Although different types of kiln are used, the principles are similar. In one type, the air in the kiln is heated by steam pipes, humidified by sprays and forced along a central floor duct by a fan near the heater. The central duct has a number of openings along its length. The conditioned air is forced out of these and it passes through the stacks of piled timber. Then it enters side ducts in the floor, to return to the heater for recirculation.

Natural seasoning is more environmentally friendly, but Kiln seasoning saves time and is therefore mostly used. By kiln seasoning methods, softwood of 50mm thickness takes approximately 3 to 5 days to reduce to 12% mc. Hardwood of 50mm thickness takes approximately double this time to reduce to 12% mc.

## Combined air and kiln seasoning

By this method, timber is air seasoned to about 20% mc, then kiln-dried for a relatively short period to achieve the required moisture content. This usually reduces the time in the kiln by approximately one third.

# TIMBER CONVERSION AND DEFECTS

As mentioned above, logs are cut (converted) into various-sized timbers before seasoning, using several different cutting arrangements. These well-established arrangements are governed either by economical consideration and waste avoidance, or a need for quality timbers with regard to the position of the annual rings and (with certain hardwood species) the position of the medullary rays. The different cutting arrangements used for log conversion include:

## Through-and-through sawn

*Figures 1.2(a)(b)*: As illustrated, the log is sawn vertically (or horizontally, depending on the machine used) with parallel cuts referred to as *through-and-through sawn, flat sawn, plain sawn* or *slash sawn*. Also referred to in the USA as *bastard sawn*. This is the most economical conversion, with very little waste. But by this method, about two-thirds of the boards have annual

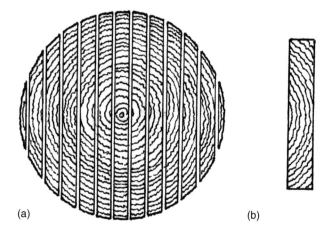

Figure 1.2 **(a)** Through-and-through sawn log (shown reassembled); and **(b)** Example of annual rings being tangential to the face of the board.

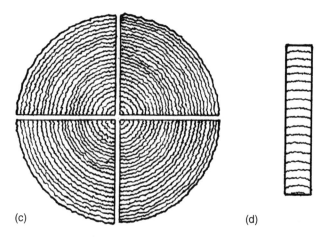

Figure 1.2 **(c)** Quarter-sawn log (shown reassembled); and **(d)** Example of annual rings being at right angles to the widest face of the board.

rings that are tangential to the face of the boards (as seen in Figure 1.2(b)) and therefore are very likely to be adversely affected by shrinkage and *cupping* (as illustrated in Figure 1.2(g)).

## Quarter sawn

*Figures 1.2(c)(d)*: As the terminology suggests, the log is first sawn into quarters (four quadrant shapes) and each of these can be converted by many different sawing arrangements. This is done either to display the medullary ray figuring on the face of the boards, or to produce more stable boards (or rectangular sections) with annual rings that are at right angles to (or not less than 45° to) the widest face of the board or section – as illustrated at 1.2(d). Four different conversions of the basic quarter-sawn shape are illustrated in Figures 1.2(e) (f)(g).

## Rift or radial sawn

*Figure 1.2(e)*: This method of converting the quarter-sawn log (as can be seen in the illustration) is the least economical method of conversion, because it leaves wasteful wedge-shaped pieces; but it is the best way of displaying the medullary rays as a feature on the faces of certain species of hardwood boards. Also, most

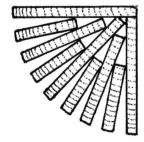

(e)

Figure 1.2 **(e)** Rift- or Radial-sawn log quadrant (shown reassembled).

importantly, because the annual growth rings are at right angles to the face of the boards (as illustrated at 1.2(d) above) there is less shrinkage across the boards' width, less warping and no cupping.

## Quarter sawn variations

*Figure 1.2(f)*: As illustrated in these three quadrants (quarter-sawn shapes), less wasteful arrangements are used, producing some boards with annual growth rings at right-angles to the face of the boards and a remainder with growth rings that are within an acceptable angle of not less than 45° to the face of the boards.

## Tangential sawn

*Figure 1.2(g)*: As illustrated, after the log is sawn into quarters, boards can also be produced with their faces tangential to the annual growth rings. This is another economical form of conversion with minimum waste, which can display the growth rings figuratively on the face of boards from such species as *pitch pine, etc*, but – as mentioned previously – tangentially sawn boards are extremely susceptible to unequal shrinkage and a form of distortion known as cupping.

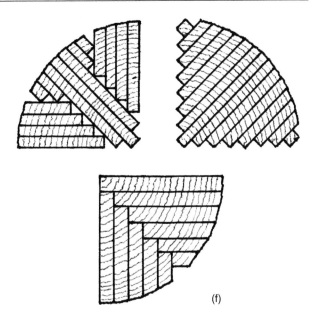

(f)

Figure 1.2 **(f)** Three less wasteful quarter-sawn variations (shown reassembled).

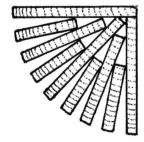

(g)

Figure 1.2 **(g)** Tangentially sawn quarter-log.

# SHRINKAGE DISTORTION

*Figure 1.2(h)*: Converted timber is subject to distortion through unequal shrinkage and it shrinks mostly in the direction of the annual rings (known as *tangential shrinkage*) – and it shrinks to a greater extent in the sapwood than the heartwood. The likely distortion this can cause is illustrated below. Note that the two concave/convex shapes are known as *cupping* – and the shrinkage effect on the square section shown on the right-hand side of the log is known as *diamonding*. With this knowledge, we can study the end grain of any piece of timber and visualize its likely distortion by knowing that the longest annual rings will shrink the most. On tangential-sawn boards, this shrinkage causes cupping.

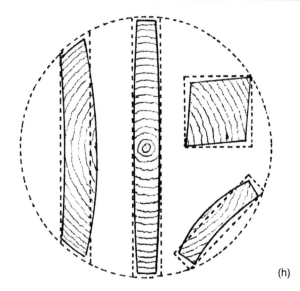

Figure 1.2 (h) Four examples of distortion through unequal shrinkage.

# DEFECTS DURING GROWTH

## Deadwood

This term refers to wood from a dead standing tree that lacks strength and its usual weight, symptomatic of the tree being felled after reaching maturity.

## Druxiness or dote

These terms refer to early decay which appears as white freckled spots or white streaks – due to fungi germs entering the tree through broken branches or other damage.

## Twisted fibres

*Figures 1.3(a)(b)*: This is recognized when the reasonably parallel lengthwise grain or fibres of a piece of converted timber are seen to slope off within the width or thickness of the timber. This is caused by straight cuts through a distorted tree trunk. Certain instances of this, as illustrated at (a), can cause *short*

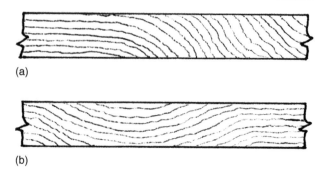

Figure 1.3 (a) Weak short grain (on R/H side) created by twisted fibres; and (b) Serpentine-shaped fibre twists that can create planing problems.

*grain*, which may seriously weaken that part of the timber and make it unsuitable structurally. Also, as illustrated at (b), if a length of sawn timber contains a *few* grain-twists (especially close together), causing it to take on a serpentine shape, it is difficult to plane the affected surfaces from either direction without being *against the grain*.

## Cup shakes or ring shakes

*Figure 1.3(c)*: As illustrated, these are segmental-shaped splits – some wide, others narrow; some short, others long – between the annual rings, reckoned to be caused by the sap freezing in the early spring. After log conversion, these concealed shakes are transferred to the surfaces or edges of the sawn stuff and can present themselves as pointed pieces of spear-shaped fibres that spring up dangerously during machine- or hand-planing operations. (As a C&J apprentice, whilst manoeuvring a room door that I was hanging and had just *shot in* (planed its edges), I had the painful experience of being speared in the fleshy part of the hand between the forefinger and the thumb by a long, wedge-shaped, tapered splinter that sprung up unseen from an arris edge.)

## Heart shakes and star shakes

*Figures 1.3(d)(e)*: As illustrated, these can be single or multiple splits that radiate from the pith (heart

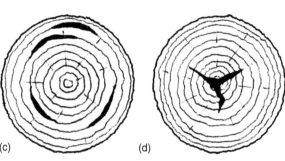
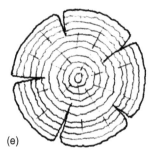

Figure 1.3 (c) Exaggerated appearance of cup or ring shakes; (d) Heart shakes; (e) Star, radial, or circumferential shakes.

shakes), or towards the pith from the circumference (star shakes). They are caused by shrinkage in an over-mature tree.

## Live knots and dead knots

Knots in converted timber are the remaining roots of the tree's branches. Small, so-called *live knots* are acceptable in commercial grading of joinery timber – but large live knots are unacceptable in good-class joinery. Knots that are loose, broken or decayed are known as *dead knots* and are unacceptable in any class of joinery. An ominous black ring around a knot is usually a sign that it is likely to fall out when the moisture content of the timber changes.

# DEFECTS DURING SEASONING

## Checks or splits

So-called *end-checks* or *end-splits* – usually of a very short length – in the ends of seasoned timbers are caused by unequal drying. When ordering timber for a particular job, it is wise to add at least 100mm for end-check wastage. *Internal-checks or splits* – usually short in length – are caused by errors in kilning temperatures. Such errors are also responsible for occasionally found *surface-checks* or *splits* on the mid-area face or edge of kiln-seasoned timbers.

# DEFECTS AFTER SEASONING

## Wane

This is caused by the log being too economically converted, leaving part of the log's round shape and inner bark (bast) on the edges of boards; producing a so-called *waney edge*.

## Bowing

*Figure 1.4(a)*: *Bow, bowed or bowing* in timber terminology refers to a segmental-shaped bend or warp in the length of a board or section of timber, which springs from the wide face of the material.

## Cupping

*Figure 1.4(b)*: *Cup, cupped or cupping* – covered previously, but shown here in 3D form – refers to a concave

Figure 1.4 **(a)** Exaggerated example of a bowed board; and **(b)** Exaggerated example of cupping.

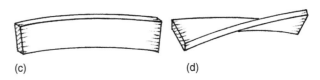

Figure 1.4 **(c)** Exaggerated example of sprung edges; and **(d)** Exaggerated example of a twisted board.

or convex distortion across the face of a board, usually caused by the board's face being tangential to the annual growth rings.

## Springing

*Figure 1.4(c)*: *Spring, sprung* or *springing* refers to a segmental-shaped bend or warp in the length of a board or section of timber, which springs from the narrow edge of the material.

## Twisting

*Figure 1.4(d)*: *Twisting, twist* or *twisted* can refer either to a distortion in the length of a piece of timber, whereby it has developed a spiral-like propeller-shape; or, in joinery terms, it refers to distortion of a framed-up unit such as a door, caused by one or more of the framed stiles or rails being twisted – or by ill-made corner joints.

## Warping

*Warp, warped* or *warping* refers to any of the above four timber defects.

# TIMBER GRADING

Although the grading of timber (according to defects, strength and appearance) varies to some extent between the different countries supplying it to bulk purchasers, it is mainly sold in six grades – although very often the first three or four grades are not separated and are sold as u/s (unsorted) timber. The difference in grades is listed below:

1. Free from all defects and with straight grain.
2. Having small, live knots only (no larger than 12mm diameter).
3. Having more knots of larger diameter.
4. Having knots, shakes and waney edges.
5. Extensive knots, shakes, waney edges and irregular grain.
6. Having all kinds of defects. Usually pulped for various uses, i.e. paper.

# 2
# Drawings and rods
## Setting and marking out

## INTRODUCTION

Most woodworkers have to be able to read scaled technical drawings, interpret them and produce work accordingly; but joiners also have to be skilled at producing drawings themselves. Apart from designing joinery items – which individual joiners might do – to scaled, drawing-board size, it is often necessary to redraw the plan- and/or sectional-views of a scaled drawing to a full-size scale (known as *setting out*) and it is also necessary to mark joint-lines and mortises, etc, onto the joinery components (and this is known as *marking out*). This additional skill, stemming from a basic knowledge of drawing principles involving sectional views, etc (to be covered here), usually develops with experience and an acquired knowledge of joinery detail.

## SETTING OUT AND MARKING OUT

Setting out and marking out, therefore, is the critical part of transferring the details and measurements from scaled drawings by others (or from one's own designs), onto the separate parts of the joinery-item – or onto separate setting out boards. This transference of detail – which has to be very precise – is widely known as *setting out* and *marking out*. The difference between these two procedures is that if a joinery item is repetitive (might need to be repeatedly produced over a period of time); or complex in detail (like a quarter-turn of four tapered (winding) steps), these items are *set out* to full-size measurements on rigid, purpose-made 'drawing boards' known in the trade as *rods*. A rod can be a 150 × 25mm prepared softwood board, long enough to accommodate the full-size sectional, detailed drawing of a door, or it can be a half-sheet (1.2m × 1.2m) of hardboard, large enough to set out the tapered steps referred to above. To emphasise the pencilled detail on rods, they are usually painted with white emulsion – and

repainted once the rod has served its purpose and is required for a different *setting out*. By comparison, *marking out* refers to one-off items of joinery, whereby the setting out of shoulders, mortises, lengths and widths, etc, is done directly onto the joinery components themselves, instead of onto a rod – and also refers to joinery components being laid onto a rod, so that the shoulders, mortises, lengths and widths, etc, can be transferred onto them. Exceptionally, if more than one identical joinery-item is being made at the same time, instead of drawing a rod, marking out is done on the separate parts of one of the items, then the various parts of this are used as 'rods' (or *patterns*) for the others. To avoid confusion, though, each first-marked part should have 'ROD' (or *pattern*) pencil-marked on it. This technique is also used by many joiners for one-off pieces of joinery, to avoid the additional work of setting out a rod.

## DRAWING PRACTICE GUIDE

The drawing practices referred to below are based on the recommendations laid down by the British Standards Institution, in their publications entitled *Construction drawing practice*, BS 1192. Note that BS 1192: Part 5: 1990, which is not referred to here, is a guide for the structuring of computer graphic information.

## COMMON SCALES USED ON DRAWINGS

*Figure 2.1*: Parts of metric scale rules, graduated in millimetres (mm), are illustrated below. Each scale represents a ratio of one unit (a millimetre) to (or *equalling*) a number of units (millimetres). Commonly used scales are 1:100, 1:50, 1:20, 1:10, 1:5 and 1:1 (full size). For example, scale 1:10 = one-tenth of full

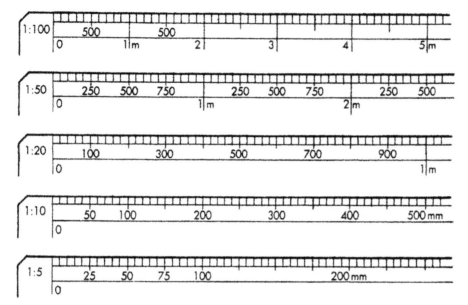

**Figure 2.1** Commonly used metric scales 1:5, 1:10, 1:20, 1:50 and 1:100.

size, or this can be thought of as 1mm on the drawing equalling 10mm in reality.

Although a scale rule is useful when reading drawings, because of the dimensional instability of paper, preference should always be given to any written dimensions displayed on the drawing.

Finally, scaled drawings not only minimize a designer/draughtsman's project to fit commercially-sized, manageable sheets of paper, they also serve to convey an important sense of proportion; i.e. if a scaled project looks disproportionate on paper, then common sense should tell us that the full-size completed project will look disproportionate.

## EXPRESSING DIMENSIONS CORRECTLY

The abbreviated unit symbol for metres is a small (lower case) letter m; and letters mm for millimetres. Symbols are not finalized by a full stop and do not use the letter 's' for the plural. Confusion occurs when, for example, 4½ metres is written as 4.500 mm – which means, by virtue of the decimal point in relation to the unit symbol (mm): 4½ mm! To express 4½ metres, it should have been written as 4500mm, 4.5m, 4.50m, or 4.500m. Also, either one symbol or the other (m or mm) should be used throughout on drawings; they should not be mixed. Normally, whole numbers should indicate millimetres; and decimalized numbers, to three places of decimals, should indicate metres. Contrary to what seems to be taught in schools, the construction industry in the UK does not use

centimetres. All references to measurement are made in millimetres and/or metres; i.e. 5 cm should always be expressed as 50 mm.

## DIMENSIONING SEQUENCE

*Figure 2.2*: The recommended *dimensioning sequence* is illustrated below. Length should always be given (or written) first, width second and thickness third. For example, a particular length of timber might be 1500 × 150 × 25 mm. However, if a different sequence is used, it should be consistent throughout.

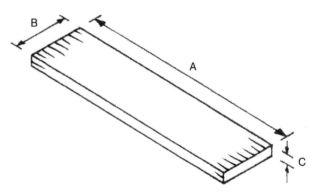

**Figure 2.2** Dimensioning sequence should be A × B × C.

# DIMENSION LINES AND FIGURES

*Figures 2.3(a)(b)(c)*: Dimension lines with open arrowheads, illustrated below at (a), are meant to be used for basic/modular (unfinished) distances, spaces or components. The more commonly used dimension lines, illustrated at (b), with solid arrow heads, are used for finished work sizes. All dimension figures should be written above and along the line; figures on vertical lines should be written, as shown at (c), to be read from the right-hand side.

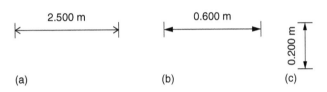

**Figure 2.3 (a)** Open arrow-heads for unfinished work **(b)** Solid arrow-heads for finished work and **(c)** Measurement read from the right-hand side.

# SPECIAL-PURPOSE LINES

*Figures 2.4(a)(b)(c)(d)(e)*: So-called *section lines* seen on drawings indicate imaginary *cutting planes* drawn at a particular point through the plan- or elevational-views, to indicate that a detailed, exposed view at this point is shown separately. The views of such exposures are called *sectional views* and are lettered A-A, B-B and so on, according to the number of sections to be shown. It is important to remember that the arrows indicate the direction of view to be shown on a separate section drawing.

# HIDDEN DETAIL

*Figure 2.5*: Hidden detail, or work to be removed, is sometimes indicated by a broken line. This is usually done on graphical drawings, especially in textbooks.

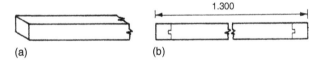

**Figure 2.5** A broken line indicates hidden detail – or work to be removed.

# BREAK LINES

*Figures 2.6(a)(b)*: End break-lines at (a), symbolized by a zigzag pattern, indicate that an object is not fully drawn. Central break-lines, as shown at (b), indicate that the object is not drawn to scale in length, even though the true measurement is given.

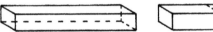

**Figure 2.6 (a)** End break-lines **(b)** Central break-lines.

# CENTRE LINES

*Figure 2.7*: Centre- or axial-lines on drawings are indicated by a thin dot-dash chain.

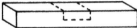

**Figure 2.7** Centre- or axial-line.

# ORTHOGRAPHIC PROJECTION

Orthography is a Latin/Greek-derived word meaning *correct spelling* or *writing*. In technical drawing the word is used to mean *correct drawing*. Orthographic projection, therefore, refers to a conventional drawing method used to display the correct layout of the

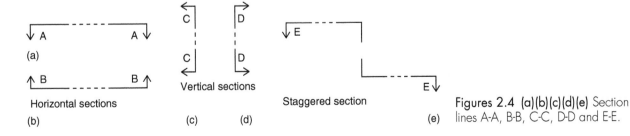

Horizontal sections

Vertical sections

Staggered section

**Figures 2.4 (a)(b)(c)(d)(e)** Section lines A-A, B-B, C-C, D-D and E-E.

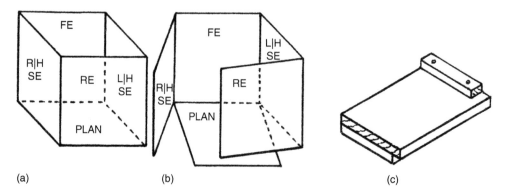

**Figure 2.8 (a)** The topless box above is used to explain the theory of orthographic 'first-angle projection'; FE = front elevation; R/H SE = right-hand side elevation; L/H SE = left-hand side elevation; RE = rear elevation; and PLAN is the view from above; **(b)** Four dihedral seams of the box are theoretically split to open it up and create one surface; **(c)** The three-dimensional object used in the illustration at 2.8**(d)** below.

three-dimensional views (length, width and height) of drawn-objects as they should be displayed separately in relation to each other on a drawing.

The two main methods of orthographic projection are 1) *First-angle* (or *European*) *projection* for construction drawings and 2) *Third-angle* (or *American*) *projection* for engineering drawings. Only method No.1 is covered here.

## First-angle projection

*Figures 2.8(a)(b)(c)*: The open, topless box illustrated in Figure 2.8(a) is used as a means of explaining the principles of first-angle projection. If you can imagine the three-dimensional object (a joiner's handmade bench hook) shown in Figure 2. 8(c) to be suspended in the box, with enough space for you to walk around it, then by looking squarely at the object from all four sides and from above, the views seen against the surfaces in the background, would be in their correct position orthographically when you open-up and lay out the sides and base of the box like a drawing.

## Laying out the box

*Figures 2.8(d)(e)*: As shown at (d), the topless box is opened out to give the views as you saw them in the box and as they *should* be laid out on a drawing. This layout is not always followed, because views are often separated onto different drawings and become unrelated orthographically. Mainly for this reason, descriptive captions are used, resulting in 'PLAN', 'FRONT ELEVATION' and 'SIDE ELEVATION', etc, being written below (or in a title box) of most drawings.

Figure 2.8(e) shows the BS (British Standards) symbol recommended for display on construction drawings to indicate that *first-angle projection* has been used.

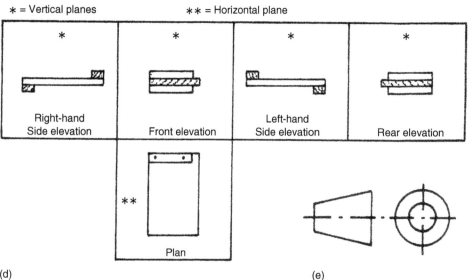

Figure 2.8 **(d)** Box laid out orthographically **(e)** First-angle projection symbol.

## Pictorial Projections

*Figures 2.9(a)(b)(c)(d)*: Other forms of orthographic projection are known as *pictorial projections*. These preserve the three-dimensional view of the drawn-object, but have a limited value in the makeup of working drawings. However, the 3-D value is very useful for illustrations and for making explanatory freehand sketches more easily understood. Figure 2.8(c) above (of the bench-hook sawing block) is a pictorial view known as an *isometric projection*. It is repeated below at 2.9(a) for comparison and is formed with 30° and 90° angles to the horizontal plane. Even when graphically explaining technical points to somebody, using a pictorial projection to guesswork angles in freehand sketches, will serve well in illustrating your point of explanation.

There are three other forms of pictorial projection, all of which come under the heading of *oblique projections*. First, at Figure 2.9(b) is an *oblique cavalier projection* of the bench hook; which requires true-shaped front elevations and true-sized side- and plan-views drawn at 45°. By this method, objects tend to look disproportionate. The second, at Figure 2.9(c) is an *oblique cabinet projection*, which is made up of true-shaped front elevations with its side- and plan-views at 45°, but scaled to only half their true depth. This half-depth technique makes the object look strangely true and proportionate. And for this reason, it rates second place to the isometric projection. The third, at Figure 2.9(d) is an *oblique planometric projection* (often

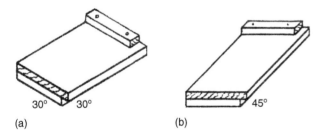

(a)                                    (b)

Figure 2.9 (a) Isometric projection (b) Oblique (cavalier) projection.

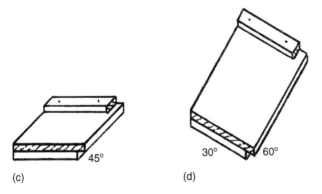

(c)                              (d)

Figure 2.9 (c) Oblique (cabinet) projection (d) Oblique (planometric) projection.

wrongly referred to as *axonometric*), which has its front elevation drawn at 30° and its side elevation at 60°; this produces a true-shaped plan view.

## MARKING AND SETTING OUT TOOLS

These vary according to what item is being marked- or set-out and they include the following, ranging from most-used to least-used:

- A sharp, ordinary *HB or 2H pencil* for general marking and setting out. The pencil-point can be conically pointed, or chisel-pointed.
- A sharp, chisel-pointed, *medium-grade leaded carpenter's pencil*, if preferred to a marking knife for marking shoulder lines; note that joiners usually sharpen pencils with a sharp, bevel-edged chisel, safely pointed away from their hands and body, whilst the pencil rests in the crotch of their free hand's thumb and forefinger.
- A sharp *marking knife*, for marking shoulder lines (especially on hardwood and quality-softwood jobs).
- A large *try square* with a 230mm blade, for squaring shoulder lines, etc, on the joinery components or for square-marking from the face-edge of a rod.
- An adjustable *combination mitre square* with a 300mm blade; note that by holding a pencil's point against the end of the adjusted blade, with the pencil and blade-end supported by the middle finger of the hand not holding the square's stock, this tool can be used as a pencil-gauge (pencil-liner).
- A 610 × 450mm *steel roofing square* (for use as a large try square on setting out rods); and a patented, aluminium *square-fence attachment* that is fixed to one of the square's outer edges, for converting it into a more easily-used try square.
- A traditional, purpose-made hardwood *pencil-liner* (also called a *runner*), that acts as a pencil-gauge for running pencil-lines parallel to the face-edge of setting out rods.
- A good-quality *steel retractable tape rule* of at least 5m length; note that, because of the loosely-riveted metal hook protruding at the blade's end, a practical technique in using these rules is needed for precise marking or setting out. For example, if setting out a standard 1 981mm door-height section on a rod or door-stile edge, I would recommend that the hooked end be used only as an anchor, pulled against the end of the rod or stile to hold the rule taut whilst marking – and not as the start of a measurement. Then, after making the initial pencil mark to clear the hooked-end area, cumulative

measurements and pencil marks can be made at the relevant points. For example, if marking out the door stile mentioned above, the first *inaccurate* mark from the hooked end could be made at 40mm, to represent the horn allowance and the outer-edge of the top rail; the second mark could then be made at 2 021mm (40 + 1981mm), to represent the outer-edge of the bottom rail. The third mark, to represent the bottom horn, would then be made at 2 061mm (40 + 1981 + 40mm). These marks are then squared across the stile's edge. This technique not only obviates the hooked-end problem, but it avoids accumulative errors that can occur when marks are not related to one datum point. The position of any intermediate rails, or lock rail, could also be marked this way – or they could be added after the extreme marks are made, by using a flat-lying four-fold rule (listed below).

- Two *stainless-steel rules*; one of 300mm and the other of 600mm length (or longer). These flat, rigid, engineering-quality rules are ideal for short measurements on marking and setting out and for setting up marking- and mortise-gauges, etc. They also make excellent, short straightedges.
- An optional boxwood- or plastic-*folding rule* (a so-called *four-fold rule*) of 1m length can be used if preferred to a steel, retractable tape rule.
- A small pair of *pencil-compasses* is occasionally useful.
- A large pair of sharp-pointed *carpenter's dividers* is also useful for setting out stair strings and other divider-work – or for use as a pair of incisive compasses.
- A pair of *trammel-* or *beam compass- heads* for circular joinery work.
- *Marking gauge(s)*, *mortise gauge(s)* and a *cutting gauge* for marking out the partly-marked joinery components.
- Clear plastic *set squares* (one of 30°/60° angles, the other of 45° angles) are sometimes of use in setting out.
- *Straightedges* of various lengths may be useful occasionally, for setting out diagonal lines, etc.

When marking or setting out items of joinery, to avoid costly mistakes try to practice the old established woodworking rule which asserts wisely that you should *always check twice and cut once.*

# EXAMPLES OF SETTING OUT AND MARKING OUT JOINERY ITEMS

*Figures 2.10(a)(b)(c)(d)*: Figure 2.10(a) illustrates a simple, outline drawing of a four-panelled room-door with section lines A-A and B-B indicating where the *imaginary cuts* are made to display the setting out detail. Note that the horizontal section A-A could be positioned above or below the middle rail – and the vertical section B-B could be positioned on either side of the vertical muntins; in both cases, the same detail would be displayed. Also note that actual section lines are not always used on drawings

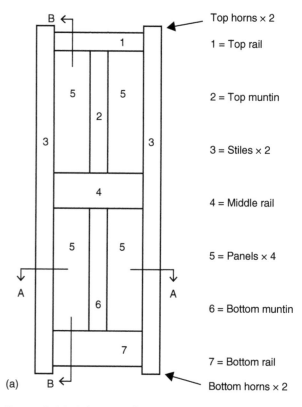

(a)

Figure 2.10 (a) Section lines AA + BB indicate where the imaginary cuts are made through a standard (1981x762mm) four-panelled door to expose the detail needed in the setting out/marking out views shown below.

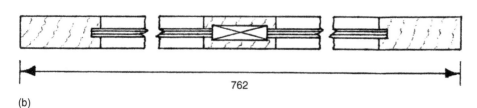

(b)

762

Figure 2.10 (b) Full-size setting out rod of section A-A, giving door-width, position of muntin, muntin-mortise (shown crossed) and shoulder lines of the three rails.

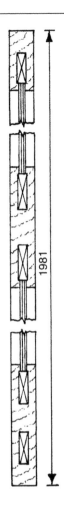

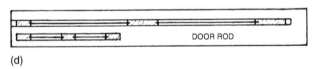

**(c)**

**Figure 2.10 (c)** Full-size setting out rod of section B-B, giving door-height, position of middle rail and positions of the five mortises required for the three rails.

**(d)**

**Figure 2.10 (d)** Outline views of sections A-A and B-B (detailed at 10(b) and (c) above), as they would be set-out on a 'door rod' of thin plywood or hardboard, etc; note that the top rail is lined-up with the left-hand stile (or vice versa); this is usually done to create speedier squaring/setting out of components with identical detail.

of joinery-items – unless designers want to highlight non-standard or complex detail – but they serve their purpose here to explain the setting out views shown at (b) and (c) above.

# PICTORIAL VIEWS OF MARKING OUT

*Figures 2.11(a)(b)(c)(d):* There are a few variations in marking out handmade joinery items – apart from those that naturally exist between handmade and machine-made products. In the latter case, if using a tenoning machine, only the end grain of one of the rails needs to be marked with the mortise gauge, so that the tenoning-block cutters can be adjusted and set to the tenon-thickness and the tenons' vertical position. My reference is to variations in marking out for handmade items. To highlight this, the pictorial views below reflect textbook marking out for hand-made joinery – which could be followed. But where I have shown dotted lines (depicting the tenon's haunch-reduction and haunch projection), I main-tain that this marking out and these cuts would best be made on the actual tenons, after they have been deeped (rip-sawn), the shoulders cut and the cheeks exposed. Note that the underside of the tenons, denoted by the dotted lines on the face-side edges, will be automatically formed by the panel-grooves (before the tenons' cheeks are exposed).

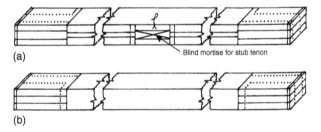

**(a)**

**(b)**

**Figure 2.11 (a)** Marking out of mortise, tenons, haunches and panel-grooves on face-side and face-edge of top rail; and **(b)** marking out on back-side and back-edge of the same rail. Note that this marked-out rail is usually G-cramped to the other unmarked rails and used as a 'rod' or 'pattern' for squaring and transferring the estab-lished marks across the face-edges. The rails are then separated to complete the marking out.

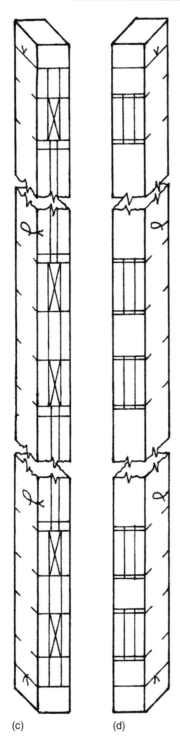

(c)          (d)

**Figure 2.11 (c)** Marking out of the horns, the five mortises, haunch-grooves and panel-grooves on the face-side and face-edge of one of the stiles; and **(d)** the marking out squared around the face-side to give the position of the mortises on the back-edge of the same stile – and the separate wedge-allowance marks, added at one-third the tenon's thickness.

# WORKING DRAWING/SETTING OUT OF DOORFRAME WITH SIDELIGHT

*Figures 2.12(a)(b)(c)(d)*: Finally, on setting out, the elevation and three designated section-views shown below should be carefully studied to help develop the skill of visualizing the separate detail related to the position of elevational section-lines (with attention to the direction of the arrow-heads).

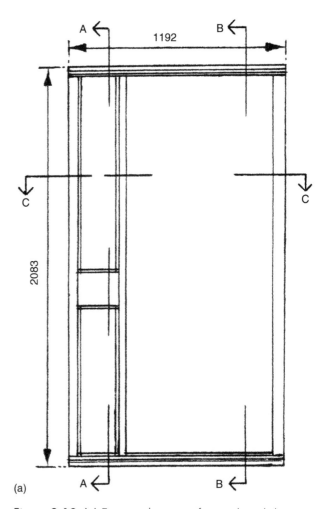

(a)

**Figure 2.12 (a)** Exterior elevation of a combined doorframe and glazed sidelight.

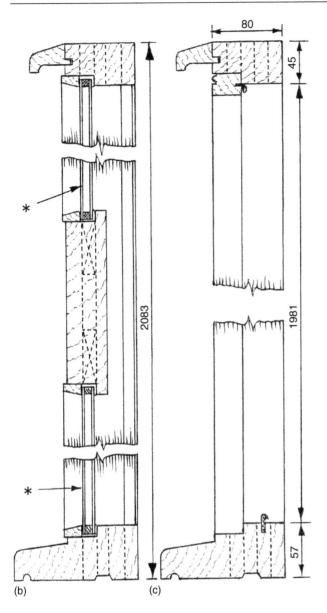

80

45

2083

1981

*

*

57

(b)                    (c)

**Figure 2.12 (b)** Section A-A through the double-glazed*(sealed-units) sidelight, showing details of frame-head with drip-mould extension, middle rail and sill, glazing beads and broken lines suggesting a jointing arrangement; and **(c)** Section B-B through the door-opening, showing similar detail, but with a bead-moulded door-stop insert at the head and a nylon water-bar insert in the grooved sill.

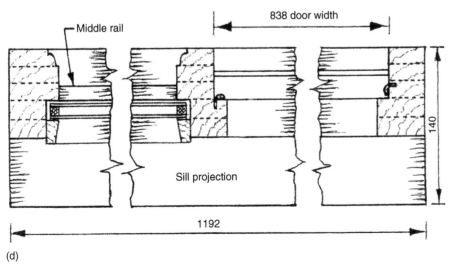

838 door width

Middle rail

140

Sill projection

1192

(d)

**Figure 2.12 (d)** Section C-C shows details of the outer jambs and the inset mullion-jamb, broken lines indicating the double mortise-and-tenon joint arrangement, sill projection, glazing beads to the double-glazed sealed-units, the middle rail, the draught-excluder inserts (the right-hand insert being on the door's hanging-side), and a plan-view of the water-bar groove.

# CUTTING LISTS

*Figure 2.13:* Cutting lists are used by joiners as a size-reference guide, or issued to wood-machinists involved in a joinery operation – and they serve an orderly function in enumerating the detailed timber-preparation of the various parts of a particular job. As an example of a typical cutting list, the components required for the above doorframe and sidelight would be laid out as follows:

| Job reference | | | | | | | |
|---|---|---|---|---|---|---|---|
| Quantity | | Sawn sizes (mm) | | | Finished sizes | | |
| | Description | Length | Width | Thickness | W | T | Remarks |
| 2 | Outer jambs | 2100 | 86 | 50 | 80 | 45 | Oak |
| 1 | Mullion jamb | 2100 | 86 | 50 | 80 | 45 | Oak |
| 1 | Head | 1200 | 86 | 50 | 80 | 45 | Oak |
| 1 | Drip mould | 1200 | 46 | 36 | 40 | 30 | Oak |
| 1 | Sill | 1200 | 146 | 63 | 140 | 57 | Oak |
| 1 | Middle rail | 320 | 205 | 50 | 198 | 45 | Oak |
| 1 | Door-stop insert | 850 | 40 | 30 | 35 | 24 | Oak |
| 2 | Glazing beads | 2400 | 25 | 18 | 20 | 12 | Oak |
| | | | | | | | |
| | | | | | | | |
| | | | | | | | |

**Figure 2.13** Cutting list for oak doorframe and sidelight.

# 3

# Joinery joints
## Their proportions and applications

## INTRODUCTION

In woodworking joinery and other related crafts such as furniture/cabinetmaking and shop-fitting, there are basically three applications of jointing timber that concern the craftsperson. *1) Frame joints*, where pieces of timber are joined together in various ways, at right- or oblique-angles to each other (as in a door with stiles and cross rails – or a drawer with front, back and sides); *2) Edge joints*, where pieces of timber are joined together side-by-side, with their fibres parallel to each other (as in the widening of a shelf or a tapered stair-tread); and *3) End joints*, where pieces of timber, with the ends of their fibres butting up to each other, are joined together (as in the lengthening of a counter- or pub bar-top, or as in modern so-called 'gluelam' work).

## FRAME JOINTS

### Mortise and tenon joints

*Figure 3.1*: This traditional frame-jointing method, which has stood the test of centuries, is still widely used in the manufacture of good-quality joinery. It has not really been satisfactorily superseded by anything more modern; although it has been challenged – more successfully in mass-produced furniture-making, than purpose-made joinery – by dowelled joints; and comb joints using improved adhesives and gluing techniques.

Possibly, one of the least-recognized features of the most common type of mortise and tenon joint (the wedged, through-tenon; illustrated in Figure 3.1), is that when the wedges are glued and driven in to the mortise on each side-edge of the tenon, the tenon (with the wedges glued to it) effectively takes on the shape and locking-action of a dovetail – that cannot easily be pulled apart. Also, because the tenon's fibres have been forced into a state of

compression by the side-pressure from the driven-wedges, any eventual shrinkage in the tenon's width will not result in lessening the holding-power of the joint. However – unlike the blunt-ended wedges illustrated below – wedges are often cut to a sharp, pointed shape, this being a deviation from the traditional wedge-shape, which was made to have a blunted point of between 2 to 4mm. The valid reason for giving the thin end of wedges this thickness was to create a state of compression deep down, near the shoulders of the tenon, as well as in the obvious upper areas where the wedges are driven in.

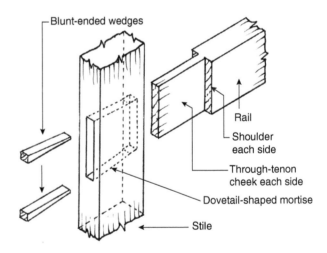

Figure 3.1 Separated view of mortise and tenon joint to plain-edged timber.

### Shoulder variations

*Figure 3.2(a)(b)(c)(d)*: As illustrated below, tenons mortised into plain-edged timber can have from one to four shoulders with which to form an abutment up against the surface of the mortised edge. And those that have none, like the ends of wooden louvres, vertical slats in a cot or garden seat, are not really tenons; they are *housing joints*.

In general joinery work, tenons normally have two shoulders, one on each side-cheek – but, occasionally, to position a mortise or a rail off centre for various reasons, a tenon may have only one shoulder and it is then known as a *bare-faced tenon*. For example, on stair work, the string-board (the *outer* string) that is tenoned into the newel posts may have bare-faced tenons with offset (off-centre) mortises in the posts – especially if the string has a minimally finished-thickness of only 28mm, which is quite common nowadays. In this case, the reason for the bare-faced tenons is to create more shoulder depth on the step-side of the string, so that the 12mm deep tread-housings (that meet the newels and run into the area for the tenons), do not cut into – and possibly weaken – any part of the 16mm-thick tenons.

Tenons with three or four shoulders (where the tenons have been reduced in width as well as the normal reduction in thickness that produces the tenon's cheeks and side-shoulders) are more commonly used in furniture-making, than joinery. This reduction in the tenon's width is done to mask the entry point into the mortise and hide any slightly-oversized mortising or mortise-chisel damage. It also eliminates the risk of the mortise's entry-point being revealed if the tenon should eventually shrink in width. Such considerations appear not to have been given to traditional – or modern – joinery. But they could be; especially on hardwood jobs.

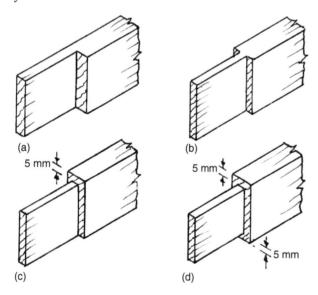

(a)

(b)

(c)

(d)

**Figure 3.2 (a)** Single-shouldered (bare-faced) tenon; **(b)** double-shouldered tenon; **(c)** triple-shouldered tenon; **(d)** quadruple-shouldered tenon.

In (c) and (d) above, where additional shoulders are created, the reduction to any tenon-edge should only

be minimal, to a maximum of 5mm (as over-reduced tenon-width reduces joint-strength) – and if the rail is not in mid-frame area, but forms a corner angle (as does the top rail of a door), then the outer edge of the tenon *must* be reduced much more to conform to the rules of tenon-reduction and haunching.

## Outer-edge tenon reductions

*Figure 3.3(a)(b)*: Framed members such as rails and stiles that form right-angles, acute or obtuse corner angles, cannot be given a full width tenon. This is because the tenon would not be contained (held in) on the outer edge. If done this way, it would, in effect, be a *corner bridle joint* (also known as an *open mortise and tenon joint*) and would not be suitable where common mortise and tenon joints were required. Therefore, to hold the tenon in a closed mortise, as opposed to an open slot, the tenon's width must be reduced. For example, the most common reduction on relatively narrow, top rails of entrance-doors is usually one third of the tenon's width. This is to allow the mortise to be reduced by the same amount so that the wedge can be safely driven-in near the outer corner. The reduction of the tenon must be enough to combat the pressure exerted on the so-called *short grain* at the end, which can shear when wedged (as illustrated in Figure 3.3(a) below). For this reason, the tenon-reduction (and short grain above the mortise) *should never be less than 38mm*; (this having been traditionally established as 1½ inches). Usually (dependant on the actual width of the top rail), the common practise of one-third reduction on wedged, through-tenons for top rails of doors violates this old joinery rule. In fact, there is lots of documentary and physical evidence that bygone scribes and craftsmen preferred even greater reduction than the 38mm rule.

It seems that well-founded woodworking rules, developed, established and passed down by

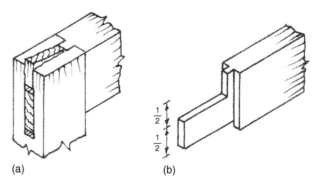

(a)

(b)

**Figure 3.3 (a)** Shearing of 'short grain' above wedged tenon **(b)** 50/50 tenon-and-haunch division.

generations of master craftsmen, have been eroded to some extent in the second half of the last century. Traditionally, a through-tenon, reduced in width to allow safe wedging near the corner (and good containment of the tenon), used to be reduced by a half (not a third, mentioned above) of the tenon's *available width* (after reduction for panel-grooves or rebates, etc), as illustrated in Figure 3.3(b). By this half-and-half rule, the structural conflict created at corners of frames, by the tenon being strong in full width and the mortise being weak, was resolved by equal proportioning of the tenon giving up half its width and the mortise losing half its length.

## Haunched areas and tenon divisions

*Figures 3.4(a) and (b):* When a portion of a tenon's width is removed, as explained above, it should not be removed entirely. A small stub, usually equal to the thickness of the tenon, is left as a projection from the shoulders. This projection is called a *haunch* and the groove to be cut in the stile to receive it is extended from the outer-end of the mortise and is called a *haunching* or *haunching groove*. The tongue-and-groove effect of the haunch-and-groove adds strength to the joint, but it also helps to restrain the tenoned rails from any future *cupping* or *twisting* movement. Note that if a stile is grooved to hold panels, the groove is used as the haunching and the haunch should fill the groove. A panel groove (if taken along the *entire* edge of stiles – which is usual) should not be wider than the tenon-and-haunch thickness, but, of course, it can be narrower.

Double tenons on the bottom rails of doors are usually found to conform to the half-and-half rule mentioned above, but – because of the necessity for a haunch at the bottom – the rule (still conforming to half the total tenon-area, half the total haunch-area) has to be in the form of four quarter-divisions: two tenons and two haunched areas, as illustrated in Figure 3.4(a).

Tenons for traditional, wide middle rails of doors (originally referred to as *lock rails*), unlike the division into thirds nowadays, also met the half-and-half rule, but in the form of two quarter-width tenons and a middle haunch area between the tenons, equal to half the tenon's total width, as illustrated at Figure 3.4(b) above. Of course, this arrangement of tenons also provided the mortises in the stiles with an adequate mid-area of long-grain for the subsequent insertion of a deep mortise lock and door-knob furniture of yesteryear.

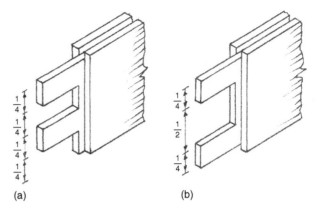

Figures 3.4 (a) and (b) Division of tenons for bottom- and middle-rails.

## Tenon-division conclusions

*Figure 3.5:* Finally, with reference to tenon-division for entrance doors (room-doors usually being mass produced nowadays), I have also seen – in text books and joinery shops – fifths used for the division of the available width of top-rail tenons, in the form of *two-fifths haunch and three-fifths tenon*. This seems to be a fair compromise between the half-and-half and the one-third/two-thirds techniques – but, mathematically, the result is so close to the *minimum of 38mm short-grain rule*, that the latter could be used without *any* dividing technique! My maths – related to the scaled drawing at Figure 3.5 below – were based on a typical, traditional top-rail width of 115mm sawn, reduced to 108mm prepared, minus a panel groove of 12mm, resulting in an *available tenon width* (ATW) of 96mm. Therefore 96 ÷ 2 = *48mm haunch* (half and half rule); 96 ÷ 3 = *32mm haunch* (one third/two thirds rule); 96 ÷ 5 = 19.2 × 2 = *38.4mm haunch* (two fifths/three fifths rule). Note that this last calculation is very close to the *minimum 38mm rule*.

Mixing the various rules, I usually favour the *fifths rule* (or *38mm minimum rule*) for top rails of entrance doors; the *thirds rule* for wide middle rails (unless a deep mortise lock is to be fitted between

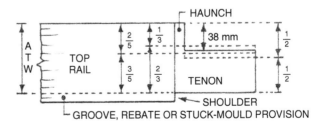

Figure 3.5 Optional divisions of the 'available tenon width' (ATW).

the tenons, then the mid-haunch area would be a half ATW, as shown in Figure 3.4(b) above); and the *fourths rule* for wide bottom rails, as shown in Figure 3.4(a).

## Tenon thickness

*Figures 3.6(a)(b)*: Again, using entrance doors as the criterion for mortise-and-tenon rules, because the tenoned rail's thickness is the same as the mortised stile, the *tenon's thickness* should normally be one third of the material's thickness. In reality, however, this rarely works out precisely, because – whether making the mortises by hand with traditional, *joiners' mortise chisels* or by machine with so-called *square, hollow chisels with integral auger bits* – the available hand- or machine-chisels are limited in their range of sizes. Therefore, in practice, the calculated one-third has to be increased or decreased to the nearest chisel size. The difference is usually no more or less than a few millimetres – and whether to increase or decrease is at the craftsperson's discretion – but it makes good structural sense to go to the nearest chisel size *above*, not below. A reduced chisel-size would mean a thinner, weaker tenon, but an increased chisel-size, whilst resulting in a slightly thicker tenon, does not significantly weaken the equally-distributed and supportive timber fibres remaining on each side of the mortise. Even so, the mortise should not be too much wider than a third.

A structural proviso to the above rule concerning doors is that the *maximum tenon thickness should not exceed 16mm*. Therefore, if a door exceeds the standard thickness of 45mm (and is over 48mm), twin tenons should be used, meaning that each *tenon's thickness* should now be one fifth of the material's thickness. Thereby, a standard 60mm-thick door should be divided by five to give twin tenons of 12mm thickness. As illustrated at Figures 3.6(a) and (b) below, the haunches can be single or separated. (I believe twin tenons should have twin haunches). This traditional *maximum 16mm rule* seems likely to be related to the comparative strength of twin-tenoned joints to single-tenoned joints when doors increase in size and weight.

## Maximum width-of-tenon rule

*Figures 3.7(a)(b)*: There is another established woodworking rule of bygone years, which prescribes that *the width across the grain of a tenon should not exceed five times its thickness* (i.e., a 16mm-thick tenon × 5 should not be wider than 80mm). This rule should be used generally, but is more relevant when tenoned rails equal the same thickness as the mortised stiles. The reason for this is that a wide, single tenon used, say, on an ex 225mm-wide bottom- or middle-rail of a door is not only adversely affected by additional shrinkage across the grain, but the greater width would also make the side-encasement material of the mortised stile relatively thin and weak on each side of the wide tenon.

Also, as illustrated in Figure 3.7(a) below, buckling and snaking of a thin, over-wide tenon can occur when the wedges are driven in. And if such a deformity develops, the compressive forces created by the wedges have changed their direction to some extent and have resulted in a sideways pressure on the thin, pliable sides of the mortise, causing them to bulge with a risk of the stile splitting beyond one or both ends of the mortised area.

Of course, common sense should prevail regarding the width of a particular tenon – and slight

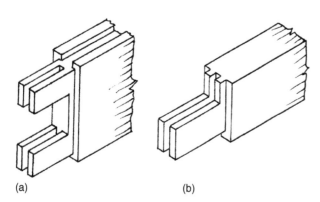

(a)                              (b)

Figure 3.6 (a) Double twin tenons on the middle rail of a 60mm-thick door (with single haunch) (b) Single twin tenons on the top rail of a 60mm-thick door (with twin haunches).

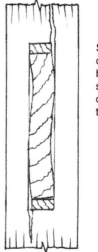

Splits in stile caused by buckling and snaking of over-wide tenon

(a)

Figure 3.7 (a) Damage caused by violation of the Width-of-Tenon Rule.

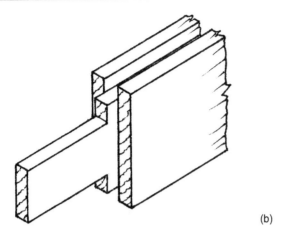

**Figure 3.7 (b)** Literal interpretation of the width-of-tenon rule.

(b)

infringements of the rule are usually disregarded. However, you do have a choice; if a basic, single tenon is not deep enough to be divided into two, after haunch allowances, yet it fails the width-of-tenon rule, then the actual tenon can be made five times the tenon's thickness, with the surplus width either reduced equally to produce a haunch on each outer edge, as indicated in Figure 3.7(b) above, to suit a mid-stile intermediate rail; or – in the case of a top-rail tenon – as well as the required reduction for the top haunch, there could also be a haunched reduction on the underside of the tenon to bring the tenon in line with the *five-times thickness rule*. In either case, it would produce a similar-looking double-haunched tenon to that shown in Figure 3.7(b).

## Haunches on traditional sashes

*Figures 3.8(a)(b)*: On traditional, ovolo-moulded and rebated sash material, because a normal haunching

groove cut in the relatively small-sectioned sash stile tends to weaken the joint, a *haunching spur* is formed from the stile's long-grain, and the end-grain of the rail's tenon is cut back into the stock to form a so-called *franked haunch*, as illustrated in Figures 3.8(a) and (b) below. A detailed and illustrated explanation of the scribing technique involved in making these joints is covered in Chapter 4.

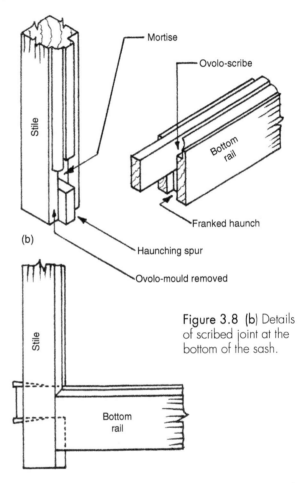

(b)

**Figure 3.8 (b)** Details of scribed joint at the bottom of the sash.

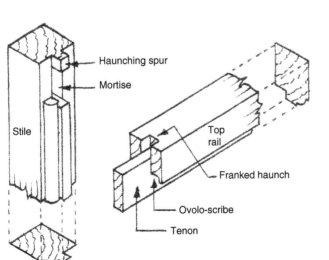

(a)

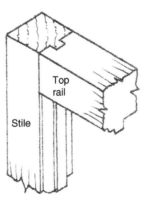

**Figure 3.8 (a)** A franked haunch and haunching spur to top joint of sash.

## Haunches on table- and chair-rails

*Figure 3.9*: Although table- and chair-making is usually in the domain of the cabinetmaker, elements of this work can bridge over into joinery – so joiners should know that the surface-appearance of haunches on rails of tables, chairs and the like is usually regarded as aesthetically undesirable; *secret (splayed) haunches* are used. An example of these is illustrated below in Figure 3.9.

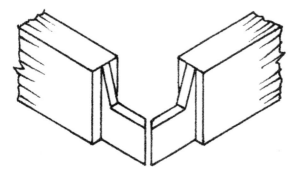

**Figure 3.9** Mitred stub tenons with secret (splayed) haunches.

## Unequal-shouldered tenon

*Figure 3.10*: This Figure shows the appearance of the haunch and staggered shoulders of a so-called unequal-shouldered tenon, when the inner edges of the stiles and rails have been rebated, but not moulded. Note that the projecting haunch equals the tenon-thickness measured from the shoulder on the rebate side (*).

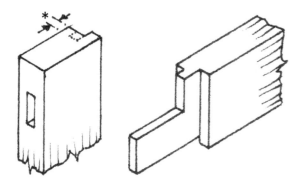

**Figure 3.10** Unequal-shouldered tenon on rebated material.

## Jointing technique for segmental-shaped inner edges

*Figure 3.11*: When a top rail of a door has a shaped inner-edge, as illustrated in Figure 3.11 below, double tenons may be required (if the *maximum-width-of-tenon rule* is violated), with obtuse-shaped shoulders to strengthen the short grain created at the base of the shaped rail. The actual angle of the short shoulder is geometrically a *normal* to the curve and its limited length is usually controlled by the depth of the panelling groove, rebate or moulded edge(s). Once this depth is set out along the short line (from prior knowledge of the edge-treatment details), this point dictates the angle of the long splayed shoulder that returns back to the edge of the stile at the top. Note that the cross-section shown below displays a so-called *raised, sunk, and fielded* panel. The bevelled border is the *raised* part; the flat, middle area is the *fielded* part, and the stepped edge around the fielded area is the *sunken* part.

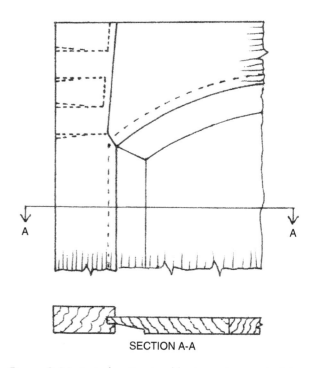

SECTION A-A

**Figure 3.11** Part elevation (and horizontal section) of the joint-arrangement for an external-type door with segmental-shaped inner edges to the top rail.

## Blind mortise and stub tenon joints

*Figures 3.12(a)(b)(c)(d)*: One of the main uses for *blind* (closed-end) mortises and stub-tenon joints in joinery is in stair work, when jointing strings and handrails to newel posts. Although such tenons are often found to be mortised into newel posts by half the post's thickness, in good joinery practise the stub tenons should be two-thirds the post's thickness. The main reason for this is that these joints (which cannot be cramped together in the usual manner) rely on wooden, so-called *draw-bore* dowels to pull-up the

shoulders – and by this process, the *short grain* created on tenons of only half newel-thickness tends to shear, making the cramping ineffective. The vertical division of the cheeks of stub tenons used for joining strings to newel posts is usually based on thirds. This gives an upper and a lower tenon and a middle haunch, as shown in Figures 3.12(a) and (b), for a stair with a step protruding past the newel post. However, when the riser-board of the first step does not protrude, but is located in the centre of the newel post, the string (as illustrated in Figure 3.12(c)) usually touches the floor and the lower tenon loses its containment. To overcome this, a haunched area of at least 38mm is allowed at the base before dividing into thirds.

## Short grain

*Figure 3.12(d)*: Note that, to simplify the mortising of newel posts, the ideal shape for oblique stub-tenons used on stair strings and handrails would be rectangular, but – although theoretically possible on string-tenons – it would create weak 'short grain' on the vulnerable triangular part of the tenon – as illustrated in Figure 3.12(d). And on handrails, there would not be enough tenon-material to create a rectangle. Another point too, from a practical point-of-view, is that newel-mortises can be cut square or sloping with the grain – but they cannot be undercut, sloping against the grain.

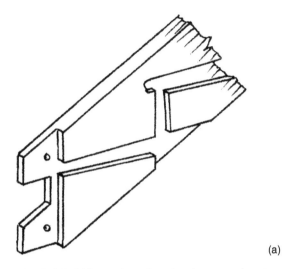

(a)

Figure 3.12 (a) Oblique, bare-faced stub-tenons showing draw-bore holes.

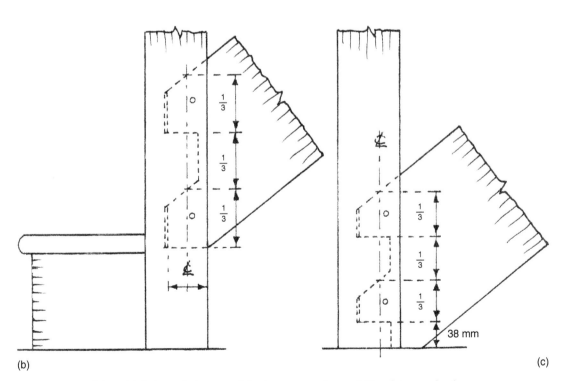

(b)

(c)

Figures 3.12 (b) and (c) Vertical division of thirds on centre-lines of ⅔ stub-tenon depth.

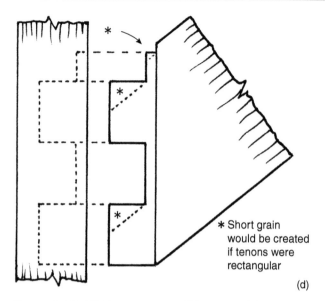

\* Short grain
would be created
if tenons were
rectangular

(d)

Figure 3.12 (d) Short grain would be created if tenons were rectangular.

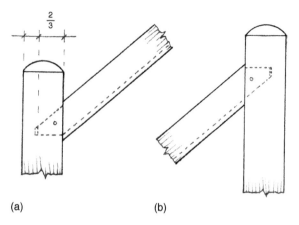

(a)                                    (b)

Figures 3.13 (a) and (b) Oblique stub-tenons into bottom- and top-newel posts.

## Handrail joints to newel posts

*Figures 3.13(a)(b)(c)(d)*: The stub-tenons on handrails, which are mortised and draw-bore dowelled into the newel posts, in a similar way to the string-tenons, are usually shouldered slightly on their top edges – as illustrated in Figures 3.13(a) and (b) below. This is done partly to mask the top of the mortise and to hide any eventual shrinkage – and partly to flatten the top of a shaped handrail. This top-edge shouldering also usually applies to handrails entering the newels at right-angles, as they do when level handrails form part of a landing balustrade – indicated in Figure 3.14(b).

Traditionally, the so-called 'sight-line shoulders' on each side of a handrail tenon (where they abut the newel post) were extended by 6mm to allow the oblique, elongated shape of the handrail to be sunk into the newel post. This provided a housing to help support the handrail in addition to the relatively thin tenon. But this practise involved squaring the 6mm acute-angled parts of handrails by chisel-paring, which demanded a high degree of skill and was very time-consuming; therefore, it is not done nowadays. However, to illustrate the practise in its simplest form, Figure 3.13(c) gives an isometric view of a portion of newel post showing a *right-angled*, blind mortise with sunken housing and draw-bore hole. This easier housing was done to receive level handrails on landing-balustrades.

Figure 3.13(d) shows the top portion of an outer string jointed to a newel post; the newel projection is known traditionally as a 'newel drop' or 'pendant', which

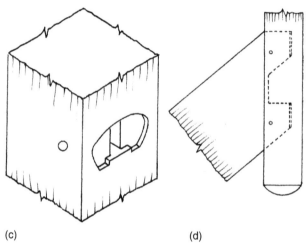

(c)                                    (d)

Figure 3.13 (c) Handrail mortise with a traditional, 6mm sunken housing (d) Underside newel-projection known as a 'newel-drop' or pendant.

was usually made to match the upper newel caps. Note that the top edge of the top tenon is not shouldered (reduced, as shown in Figure 3.2(c)) to mask the possibility of the top of the mortise showing on an inferior fit – or in the event of shrinkage across the string's width. The reason for this is that, after a stair has been fitted on site, the top edge of the outer string is fitted with a grooved capping to receive the balusters. This capping would mask any gaps that might show.

## Tenon-thickness for handrails

Although, theoretically, the handrail's tenon-thickness could be up to a third of the handrail's shaped, *mean* width, to save time it is usually made to suit a 16mm chisel size, thereby equalling the other mortises made for the string-tenons.

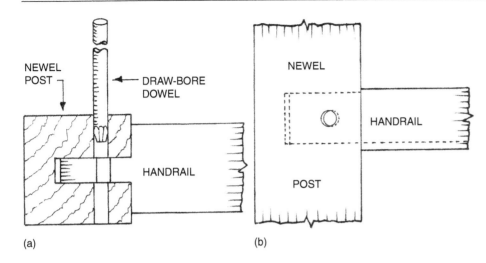

(a)                                              (b)

Figure 3.14 (a) Horizontal section through newel and handrail showing draw-bore technique; (b) Part-elevation showing hole in tenon offset by 2mm towards the hand-rail's shoulders.

## Draw-bore Technique

*Figures 3.14(a)(b)*: Dowelling used for *draw-bore dowels* or *pins* (having replaced the archaic term *treenails or trenails*) causes some confusion when reference is made to *'steel' draw-bore pins* used by joiners for initially testing the quality of the draw-bore effect. The dowels are usually of 12mm Ø (diameter) and should be the same timber species as the newel posts. However, close-grained, straw-coloured Ramin dowelling (a *hardwood* species commonly available commercially) is often used for draw-bore dowels on *softwood* newels. The dowel's centreline position should be marked on the newel posts at twice the dowel's diameter, i.e. 12 × 2 = 24mm from the shoulder. Vertically, this point should be in the centre of the available tenon-height. After being carefully drilled in the newel posts, the strings and handrails are fitted (separately) – ensuring close-fitting shoulders – and a 12mm Ø pointed auger-bit is inserted to mark the tenon. After withdrawing the tenon, a second mark is made 2mm towards the shoulder and a 12mm Ø offset hole is drilled, as indicated in Figures 3.14(a)(b).

## Mild-steel draw-bore pins

*Figure 3.15*: Because the jointing of strings and handrails to newel posts must be fitted and checked – stair-makers ideally require a pair of *mild-steel draw-bore pins* to test the draw-bore effect. Illustrated below, these pins are made from 11mm Ø mild steel and need to be about 300mm long. One end is ground to a conical taper and the other is bent over to assist in withdrawing the pins after the joints are checked.

Figure 3.15 Draw-bore pin.

## Fox-wedged stub tenons

*Figures 3.16(a)(b)*: These traditional joints deserve mentioning as, theoretically, fox-wedging is an excellent way of strengthening stub-tenon joints – effectively turning them into secret dovetails. But in reality, they are not very practicable and I have never believed that they have been used professionally to any great extent. To be successful, the size of the wedges must be carefully judged in relation to the degree of under-cut slope at the ends of the mortise. Also, the length of the wedges is critical. The marrying of the joint can so easily fail if the wedges are too long, too short, too thick or too thin. Also, when the wedges touch down on the bottom of the blind mortise and pressure is applied to close the joint and force the wedges into the saw kerfs in the tenon, the wedges sometimes move along the line of least resistance, causing them to fracture and bend over sideways (indicated in Figure 3.16(b)). This can obstruct the full entry of the tenon. Finally, there is also a risk – as with all incisive wedging that penetrates into the actual timber – that the wedging/splitting action occurring in the tenon will spread and appear in the rail itself.

# MAKING BASIC MORTISE-AND-TENON JOINTS

Having covered most of the fundamental technicalities concerning mortise-and-tenon joints, I can move on to the making of them. The detailed explanations and illustrations here will mainly be to do with the techniques involved in *the use of hand tools* (this being usually regarded as the best foundation for today's trainee mechanized-joiners, wood-machinists *and* DIY enthusiasts), but references will also be made to the use of powered portable tools and fixed

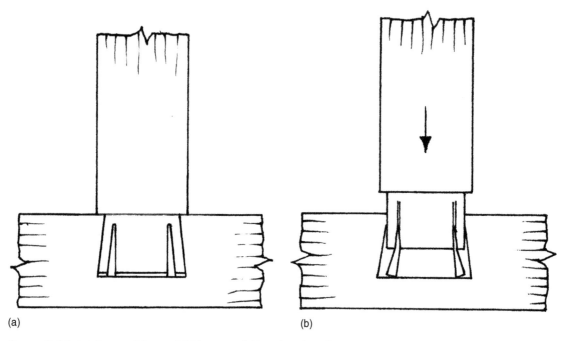

(a)                                          (b)

Figure 3.16 (a) Successful joint (b) Unsuccessful (wedges bending).

woodworking machines that can be used if available – and if the operator is competent and knowledgeable enough to use them safely.

## Through-tenons and stub-tenons

The techniques for making stub-tenons and blind mortises are very similar to those used to make wedged through-tenons, but less involved, so through tenons are explained here. Because there is a separate chapter in this book covering *setting out* and *marking out* the joints with a *try square* and *mortise gauge*, etc, (Chapter 2) the procedure explained here starts after the marking out has been done.

## Cutting the tenons

First the tenons are *deeped* (cut along the grain, through the deepest part of the rail's tenons) to produce the tenons' cheeks; and although there are various ways of doing this with machines, there is only one satisfactory way by hand. This is by sawing down the grain in a vice, on the waste-side of the tenon, tight against the gauge line. Although a modern hard-point handsaw with a universal tooth-shape would be used to do the job nowadays, being a traditionalist myself, I would prefer – in the absence of machinery – to use a genuine 660mm (26") rip saw with 5 or 6 points per 25mm (pp25mm). Such saws could be used for deeping tenons of about 100mm wide (top rails of doors) to 225mm wide (middle and bottom rails

of doors). Lesser sized tenons, down to delicate work for small cabinets, would require either a maximum-length hardpoint tenon saw of 300mm × 13 pp25mm, or – my choice – a traditional brass-backed tenon saw of 350mm. Un-mechanized joiners of yesteryear (with saw-sharpening skills) often converted the 13 pp25mm – 13 ppi (points per inch) – crosscutting teeth of one of their tenon saws to rip-cutting teeth of 7 pp25mm. This was done by filing out every other tooth with a triangular-shaped saw-file to a pitch angle of 87 to 90° at right-angles to the blade to create rip-sawing teeth. Such converted saws produce better, more accurate and faster saw cuts along the grain – and are a pleasure to use.

## Deeping variations

*Figures 3.17(a)(b)(c)(d)(e)*: There are a few variations in the method of deeping with hand saws. My way, as illustrated below, is as follows:

(a) and (b) Set the rail vertically in the vice, tilt the hardpoint- or rip-saw (whichever one of the four alternatives discussed above) towards the far top corner of the rail and start sawing on the side of the tenon that does not obscure the view of the gauge line. Bring the saw down within a few strokes to establish a shallow cut across the end grain;

(b) and (c) Keep the saw in the shallow cut, but slacken off the weight of the saw on the front starting point and start working the saw deeper on the visual nearside edge, towards the shoulder;

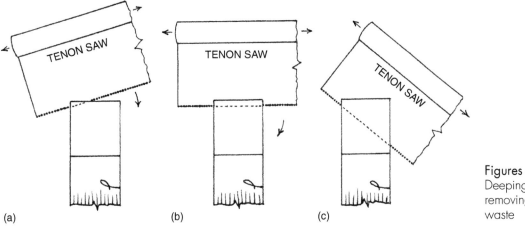

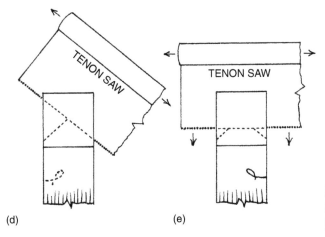

**Figures 3.17 (a)(b)(c)** Deeping procedure for removing tenon-cheek waste

(a)                    (b)                    (c)

**Figures 3.17 (d)(e)** Final stages of the deeping procedure.

(d)                    (e)

(d) Turn the work (or yourself) around and run the saw in the established cut, down again to the shoulder;

(e) Level out the saw and carefully finish sawing down to the shoulder line.

## Deeping repetition

Repeat the above procedure on the other cheek of the tenon – and on all the ends of the rails to be tenoned. Note that if the inner-edges of the rails are to be grooved, moulded or rebated, the rails are left with the tenons deeped, but not yet shouldered; the tenon-cheek waste material is left on to allow a better start and run-off of the hand planes, powered router or spindle moulder used to form the edge-detailing.

## Cutting shoulders

*Figure 3.18*: Once the edge detailing has been done, each rail can be laid on a pair of wooden bench hooks and the shoulders can be cut with a tenon-saw (not the one with the rip-saw teeth!). When the shoulders have been scored with a marking knife, it is good

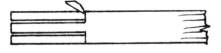

SECTION A-A THROUGH CHISEL

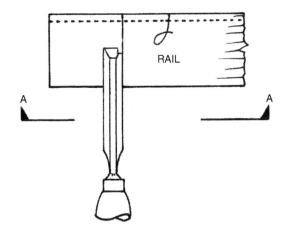

**Figure 3.18** Paring the V-shaped furrow against the tenon's shoulder.

practise to keep a fraction away (0.5 to 1mm) from the knifed shoulder and then this can be more precisely finished by vertical paring with a wide, sharp bevel-edged paring chisel.

On certain softwood jobs, where the shoulders are marked out by pencil, paring can be avoided if a good degree of tenon-saw skill has been acquired and the shoulders can be carefully finished with the tenon saw. However, if your skill is not yet developed and you need a shoulder to lean against, a small v-shaped furrow can be formed to hold the tenon saw against the shoulder. The v-shape is started with a marking knife by simply knifing a deep shoulder line vertically against a try square, then by paring the waste-side of the knife-cut across the rail (from each side) with a very sharp paring chisel tilted to about 30°, as illustrated in Figure 3.18.

## Producing wedges

*Figure 3.19*: Once the basic tenons have been formed, the reduction for the haunch-allowance can be marked onto the tenons' cheeks. Then each rail is placed upright in the vice and two or more wedges – the angles of which are usually guessed at by experienced woodworkers – are cut from the waste area before the exact width of the tenon is cut and the haunch projection formed by a tenon-saw cut across the grain. However, whether guessing or marking out the wedges as illustrated in Figure 3.19, the wedge-angle should be about 85° (1 in 12).

This traditional technique produces wedges quickly from waste material of the correct tenon-thickness and eliminates what-can-be a fiddly job on separate, short pieces of wood.

## Machined and part-machined tenons

When the machine is set up accurately, precise tenons with well-cut shoulders can be produced on heavy, fixed *single-* or *double-ended tenoning machines*. With such machines the usual practise is to mortise the stiles first. This enables a short-stub (about 6mm) trial-run-tenon to be machined on a rail-end for trying into a mortise to check for fit and flushness. Note that only the end grain at one end of the trial rail needs to be marked with the mortise-gauge, to enable the (electrically isolated) top- and bottom-cutters to be manually revolved, compared and adjusted to the gauge lines.

For less-mechanized workshops, the tenons are usually deeped on a *narrow band-saw machine*, then shouldered by hand as described above under *Cutting Shoulders*. Note that it pays to use the widest blade on your band saw – to avoid 'snaking' – and only start the cut against the machine's fence, then release it and complete the cut carefully by freehand and visual control.

## Mortising by hand

*Figure 3.20*: Although traditional hand mortise-chopping techniques were replaced many years ago in industry by a variety of machine methods (a mortising machine using hollow square mortise chisels with integral auger bits being the most popular in small-sized workshops), there are certain limitations by machine – one of them being when a mortise is required to be cut at an angle to the timber's surface. Such is the case when an oblique mortise is being cut into a newel post. A portion of it has to be chopped out by hand with a mortise chisel – two types are illustrated in Figure 3.20. Apart from this, I believe that hand mortise-chopping contains a valuable, transferable skill-element. And for that reason, it was

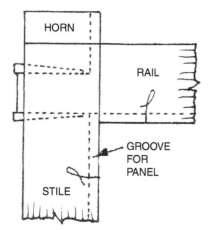
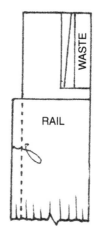
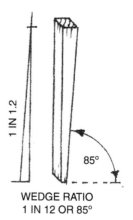

Figures 3.19 The technique of producing wedges from tenon-material.

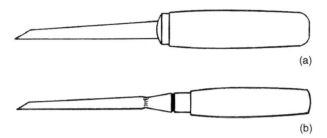

Figure 3.20 (a) A heavy-duty mortise chisel; (b) a sash mortise chisel.

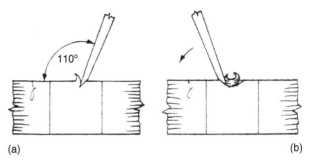

(a)                                                    (b)

Figures 3.21 (a) and (b) Initial stages of mortising technique.

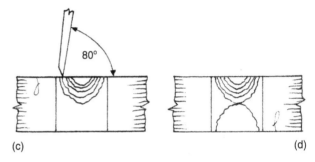

(c)                                                    (d)

Figure 3.21 (c) Chopping angle; (d) Work turned over.

taught and practised in training establishments and detailed in text books. And although hand-mortising methods vary, I believe that my innovative – but not unorthodox – way of doing it is relatively fast, efficient and accurate.

As mentioned previously for tenoning, because there is a separate chapter in this book covering *setting out and marking out* the joints with a *try square* and *mortise gauge*, etc, the procedure explained here starts after the marking out has been done. However, I will mention that the allowance for wedges, outside of the pencil-squared mortise-lines on the back edge of the stile, should be one third of the tenon-thickness. After these ⅓rd allowances are pencil-squared – usually by judgement rather than measurement – the eventual mortise-gauge lines extend each side to include them.

## Mortise-chopping

*Figures 3.21(a) to (j):* To make a good job of hand-mortising, concentration and attention must be given to sighting the verticality of the mortise chisel prior to striking a blow with a mallet. If you can achieve this with every blow and apply it to a good chopping technique, you might be surprised by how fast a good mortise can be produced. Chopping techniques vary from orthodox to unorthodox methods – and one of the latter that uses a series of drilled holes and is pared out, as opposed to being chopped and reamed out with a proper mortise chisel, should only be used on site-carpentry work, or if a mortise is wider than the largest mortise chisel available. The following is a simple variation on the orthodox, traditional method, which personally evolved many years ago. My way, as detailed and illustrated below from (a) to (j), is as follows:

(a)(b)(c) Position the mortise chisel between the gauge lines in the middle area of the mortise, leaning it at about 110° away from the flat underside of the chisel and strike a light blow. Turn the chisel around and place it at about 10mm from the first cut, leaning it the opposite way and strike another light blow. Both of these actions aim to produce a small

V-shaped incision with the short-grain fibres broken away between the cuts. Keep repeating the procedure, with more powerful mallet blows, working away from the last cuts by about 6mm on each side, widening and deepening the V-shaped mortise with each cut. The angle of chisel will need to reduce to about 80° and the V-shape should now be looking more like a U-shape. After each action the chisel is removed by levering downwards carefully to ream the sides of the mortise and remove the wood chippings. Stop the action when you are about 5mm away from the mortise lines at each end, as any leverage beyond this will cause exposed bruising. At the same time, make sure that the deeper chisel work, at the inverted apex of the mortise, has approximately reached half the full depth.

(d)(e) Turn the work over and repeat the whole procedure again. When nearing completion, you will sense a feeling and a sighting of breaking through in the depths of the middle area. When this happens, enlarge the break-through point slightly to check on the side-alignment of the upper mortise to the lower one.

(f)(g) Now hold the chisel upright in both vertical planes, positioned off centre in the deep part of the mortise and start chopping estimated mid-depth cuts, in about 6mm steps, with the flat side of the chisel working towards yourself and the finishing line of the mortise. Repeat this procedure on each side of the mortise.

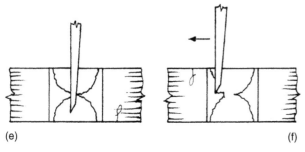

**Figure 3.21 (e)** Breaking through; **(f)** Chopping to the left.

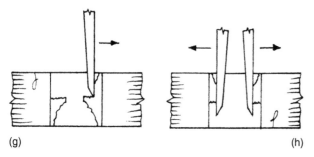

**Figure 3.21 (g)** Chopping to the right; **(h)** Work turned over.

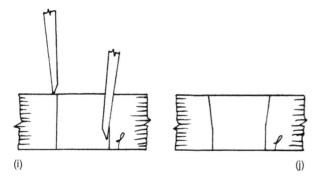

**Figure 3.21 (i)** Creating wedge-slopes; **(j)** Completed mortise.

(h) Now turn the work over and repeat the procedure again. But this time, let the chisel go partly through to the cleared area below to promote alignment.

(i)(j) Finally, with the back edge of the work uppermost, place the chisel on the wedge marks and chop the slopes so that they break into the mortises at a half to two-thirds of the overall depth. Once you have gained experience and confidence, try to achieve this with one mallet blow per wedge slope.

## Holding the work whilst mortising

When mortising, the work should be resting on a solid surface and be held upright. It can be cramped to the top edge of a bench in various ways. For example, a short offcut of timber can be fixed upright in a vice, with the adjacent work G-cramped to it on the edge of the bench – or the work can be held on the edge of the bench by means of a sash cramp fixed vertically to the underside edge of the bench's side-rail. If large enough (such as a doorway door-stile), the work can also rest on one or two purpose-built wooden stools. This enables the craftsperson to sit on the work and adopt one of the best hand-mortising positions possible.

## Purpose-built mortising stool(s)

*Figure 3.22*: Mortising stools, like traditional saw stools, were made by the craftsperson themselves and therefore varied in design – often depending on what timber was available at the time. For those hand-techniques' enthusiasts interested in following and emulating past master-craftsmen's methods, I have designed one of these stools, illustrated in Figure 3.22, which has the advantage of being able to hold lighter work as well, via a G-cramp (or cramps) being fixed to the underside of the stool's relatively thin top.

## The stool's specification

My design uses 50 × 100mm prepared softwood for the top, the legs and the T-shaped feet. However, traditionally, the top was made more solid by using 75 × 100mm or 100 × 100mm prepared. The tread boards,

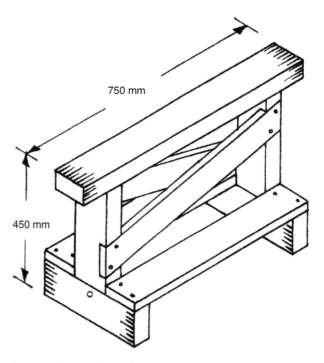

**Figure 3.22** Traditional-type mortising stool.

acting as stretchers, can be 25 × 100mm sawn or prepared and the braces can be 25 × 75mm prepared. The legs can be stub-tenoned and draw-bore dowelled (with 12mm Ø dowels) into the T-shaped feet and stub-tenoned into the top. The tread boards can be screwed or nailed and the diagonal cross-braces should be housed in and screwed or nailed. The length, height and width of the stool can be varied to suit yourself.

## Machine mortising

*Figure 3.23*: As mentioned under *Mortising by Hand*, machine mortising in small workshops is usually done with a machine using a chain cutter and/or a chisel-mortising facility. The principle involved with the latter, using square hollow mortise chisels with integral auger bits is that whilst the revolving auger cuts away the major part of the material – slightly in advance of the four-sided chisel – the closely-positioned, sharp edges of the square hollow chisel cuts away the corner fibres with a paring action generated by the hand-operated lever. In a well-sharpened tool, these chipped fibres are broken up small enough to pass up the revolving flutes of the auger and are ejected through an elongated aperture in the *side* of the chisel – not the *front*. The following machining points must be taken into account:

- When setting up a chisel into a mortiser, always allow 1mm clearance between the chisel's sharpened inner-edges and the upper surface of the auger-bit's larger diameter spurred-cutter. This can be achieved by fixing the auger bit (whilst held

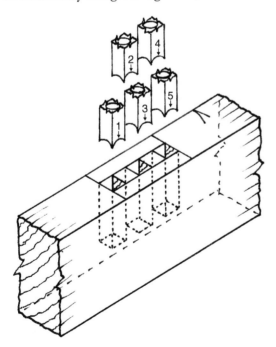

Figure 3.23  1,3,5,4,2 sequence of mortising.

tight-up within the chisel) into the chuck – initially therefore without any allowance – but with a one-penny coin placed between the chisel's shoulder and the underside of the chuck. The clearance is obtained (after fixing the auger bit) by removing the coin and re-fixing the chisel tight-up to the underside of the chuck. This method eliminates guesswork.

- Now pull down the lever to lower and position the chisel adjacent to the machine's fence, then slacken and re-fix the chisel (tight-up to the underside of the chuck again!) after squaring it to the fence with a small, metal try-square.
- If mortising for through-tenons, set the vertical stops on the mortiser to achieve an approximate ⅝ depth of mortise, so that the components can be mortised on both edges (face-edges and rear-edges), with the face-side marks *always* up against the machine's fence.
- Finally, try to avoid progressive, side-by-side consecutive mortising, as this can put a strain on the chisel. This is because, with a completed square hole at one side only, the chisel tends to strain towards the side with least resistance. However, there is no strict inconsecutive hit-and-miss sequence. For example, if mortising a 76mm wide mortise with a 16mm square chisel, at least 5 plunges will be required. Numbering these as 1 to 5 from left to right, the ⅝ depth mortising sequence could be 1,5,3,2,4; 5,1,3,4,2; 1,3,5,4,2, etc, as illustrated below. And, of course, apart from plunges 1 and 5, which have to be carefully related to the pencil-marked position of the mortise, the position of plunges 2, 3 and 4 is not critical and are only guesstimated.
- After being released from the mortising machine, the slopes of the pre-marked wedge positions can be cut by hand with a firmer- or mortise-chisel, in a similar way to that detailed in Figure 3.21(i) and (j).

# DOVETAIL JOINTS

## Setting out

*Figure 3.24*: There is a traditional method for setting out multiples (sets) of dovetails with a rule – which is illustrated and explained here – but many years ago I devised an alternative technique involving the use of a small pair of dividers. Possibly other woodworkers, with a love of practical geometry, may also have come up with this idea, but not to my knowledge. However, if making sets of dovetails by hand, I recommend the divider method for its simplicity, accuracy and speed.

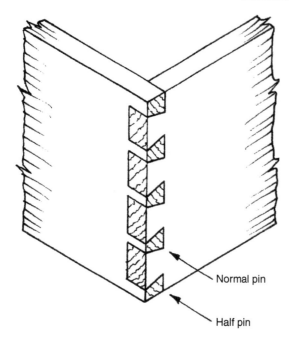

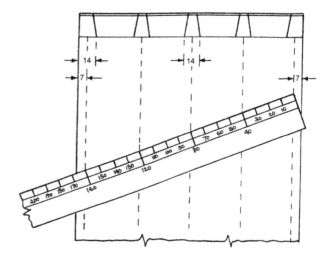

Figure 3.25 Traditional method of setting out dovetails.

Figure 3.24 A set of four through-dovetails.

The technique can be applied to setting out either *through-dovetails* or *lapped-dovetails*.

## Dovetailing terminology

- *Dovetails*: The double-slope shapes on the face-ends of the material, likened to the appearance of a dove's fan-shaped tail
- *Dovetail sockets*: The double-slope recesses on the end grain of the material, cut away to receive the dovetails
- *Normal pins*: The double-slope, inverted dovetail shapes separating the sockets and locking (pinning) the dovetails into them
- *Half pins*: The single-slope, inverted half-dovetail shapes positioned on the outer edges of the timber (note that they are always referred to as *half pins*, even though they are usually larger than an exact half division of a normal double-slope pin)
- *Pin sockets*: The double-slope, inverted dovetail recesses between the dovetails, cut away to receive the pins.

## Traditional method

*Figure 3.25*: The traditional technique of setting out sets of dovetails with a rule also conforms to a simple method of geometry. Such a method avoids mathematical division of the actual width of the timber to be divided for a number of dovetails and adopts suitable non-fractionized units that, when multiplied by the chosen number of dovetails, must always be greater than the actual width of the timber. As illustrated in Figure 3.25, the number of oversized units (multiples of 40mm in the example) shown on the rule is equal to – and decided by – the number of dovetails you choose to use on a particular width of timber. However, bear in mind that the dovetails – for shrinkage reasons – should not be wider than 38mm at their base. With repetition and experience, the amount of dovetails required can usually be visualized.

By this method, once the shoulder lines (equal to the wood's thickness + 1mm for initial end-grain planing and eventual *cleaning up*) have been pencil-marked squarely around the timber to be dovetailed, you must first decide what size the widest part of the normal pins will be. It is wise to make them slightly larger than one of your bevel-edged chisels. This will simplify the eventual vertical paring of the waste material against the shoulder line of the pin sockets. In this example, I have chosen 14mm to suit a 12mm (or ½") chisel. Then halve this chosen measurement and pencil-gauge a 7mm (or ¼") line along the face-side of each outer edge, as illustrated.

Next, lay a rule across these outer-edge lines and pivot it as shown until the multiple of your chosen units (40mm × 4 (for number of dovetails) = 160mm) is shown between the lines. Carefully mark the three points given at 40mm, 80mm and 120mm, as these will be the centres of the normal pins. Produce pencil-gauged lines from these three points up to the shoulder line, then mark 7mm points on each side of these lines from which to mark the slopes of the pins and tails.

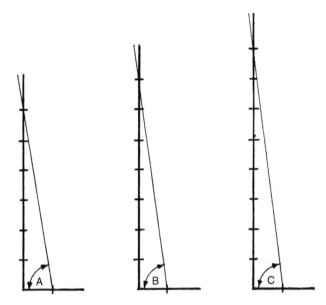

Figure 3.26 **(A)** 1 in 6 softwood angle; **(B)** 1 in 7 medium-density wood angle; **(C)** 1 in 8 hardwood angle.

## Dovetail Slope

*Figure 3.26*: The dovetail angle for softwood is usually set to a ratio of 1 in 6. This can be set up on a sliding bevel by alignment to the angle formed from a right-angled pencil line across a board which has been marked 6cm across (or 6 inches) and 1 cm (or 1 inch) along the edge, as illustrated. Also shown is the angle for medium-density wood (1 in 7) and hardwood, which is usually set to a ratio of 1 in 8.

## Dovetail templates

*Figures 3.27(a)(b)*: Dovetail templates made of beech or other close-grained hardwood species can be made

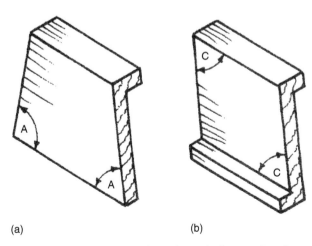

(a)                                    (b)

Figure 3.27 **(a)** Template for softwood; **(b)** Template for hardwood.

as an alternative to using a sliding bevel. Two variations in design are illustrated here. Metal templates can also be made or bought.

## Setting out dovetails with dividers

### Equal tails and pins

*Figures 3.28(a)(b)*: As with the traditional method, the shoulder lines (equal to the wood's thickness + 1mm for initial end-grain planing and eventual cleaning up) must be pencil-marked squarely around the timber and the centre line(s) established on the face side. With non-committal positioning of a pair of dividers, the centre line is used to plot the dividing points for the tails (dovetails). The semi-circles shown indicate the pivoting movement (steps) of one divider-point mark to the next. If you do not possess a pair of dividers, a pair of tautly-tensioned compasses will do. First – as with the traditional method – you must decide how many dovetails are required.

If you decide to have four dovetails across the width, as illustrated below, the dividers must be *stepped*

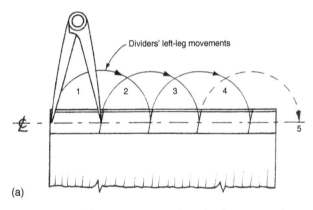

(a)

Figure 3.28 **(a)** First setting out: four divider-steps to the right.

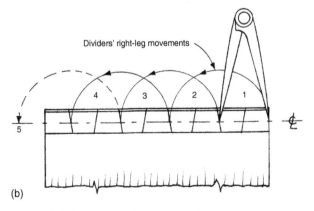

(b)

Figure 3.28 **(b)** Second setting out: four divider-steps to the left.

*out* (progressively pivoted with alternate divider-points remaining lightly pressed in the timber) by trial-and-error stepping, five steps from left to right. The fifth step, where one divider-point is off the timber, over the edge, is only needed initially to allow visual judgement to be made as to whether half-a-step, more or less, has been achieved.

If under half-a-step appears over the edge, try again after adjusting the dividers. If slightly over half-a-step, you could adjust the dividers and try again, or you could let it go. Letting it go will result in slightly wider dovetails.

When satisfied with the divider-setting, commit four steps from left to right and then four steps from right to left. Depress the leading divider-leg slightly at each step to mark the timber. With a bevel or a template, mark the alternating dovetail angles through these points and square them across the end grain. This will give tails which are equal to the shape and size of the supporting pins; but the half-pins on the outer edges tend to look oversized. Even so, it has to be said that this arrangement of equal (or near equal) tails and pins presents a structurally strong joint – and should therefore be used on certain jobs where a load has to be carried.

## Large tails with smaller pins

*Figures 3.29(a)(b)*: If small pins and larger dovetails are preferred or required, the technique of setting out is the same, but the distance marked X, shown over the edge in divider-step 5 in the illustration, must be well over half-a-step. Once again, trial-and-error stepping out will bring about whatever size tails and pins are required. After stepping out four divider steps one way, say left to right, then step out and mark four steps the other way. As before, mark the alternating angles through these points and square them across the end grain.

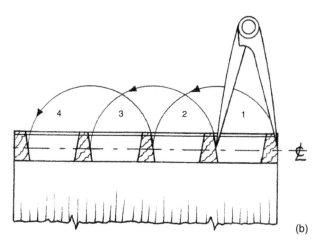

Figure 3.29 (b) Second setting out: four divider-steps to the left.

## Dividers

*Figure 3.30*: Two types of dividers are illustrated here. The spring-loaded pair, shown in Figure 3.30, is relatively cheap to buy and the other pair is hand-made from mild-steel sheet of 1mm thickness. They are 100mm long. The degree of tension required between the two legs was achieved by careful tightening of the rivet at the top.

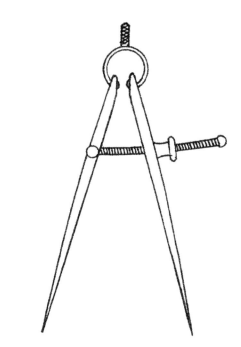

Figure 3.30 Spring-loaded dividers.

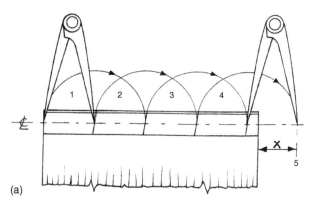

Figure 3.29 (a) First setting out: four divider-steps to the right.

## Making dovetails by hand

*Figures 3.31(a)(b)*: Joinery and cabinetmaking techniques are generally universal, but are often

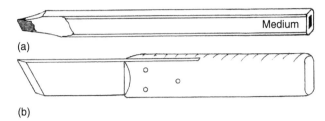

Figure 3.31 (a) A carpenter's pencil converted to a joiner's knife-like pencil (b) A typical single bevelled-edge marking knife.

personalized by individual, experienced woodworkers, resulting in interesting variations. My own slant on dovetailing techniques is as follows:

Whether the material is softwood or hardwood, I use a combination of an ordinary HB pencil (for marking face-side and face-edge marks), a carpenter's oval- or rectangular-shaped pencil (for marking the shoulder- and centre-lines, dovetails and pins) and a marking knife for the intermittent shoulder marks and (on hardwood jobs) for marking the pin-positions between the dovetails. The carpenter's pencil, as illustrated, has to be shaved well back on each wide-side to produce a very sharp (knife-like) skewed chisel-edge. The marking knife, also with a skewed cutting edge, has a sharp 20 to 25° ground and honed bevel on each side *or on one side only*. The double bevel allows for left-hand or right-hand use and suits users who prefer to lean the knife over, as I do, (against the edge of a try square) to gain better visual appraisal of being on target. However, marking knives with a single bevelled edge have the advantage of being sharper – because the ground and honed bevel is only 20 to 25°, whereas a double bevel (2 × 20 to 25°) produces a less incisive edge of 40 to 50° – and single bevelled knives (if you regrind one and have a handed pair) are ideal for laying flat and marking the pins between closely-spaced dovetails.

## Through dovetails

The dovetailing procedure outlined here is related to the making of a box with four equal-length sides measuring, say, 450 × 145 × 12mm (perhaps to be clad with 4mm plywood and then sawn continuously around the glued and cleaned-up, closed box to produce a lid). Having prepared the timber to the sectional sizes and marked the selected surfaces and edges with face-side and face-edge marks, the four pieces must be cut carefully to length + 2mm (452mm). They are now ready to be marked with joint shoulder-lines of 13mm (12mm + 1mm) to be squared around each end. The 1mm allowance is for initial end-grain planing and eventual cleaning-up.

One of the ways that I do this is to cramp the four carefully-lined-up pieces together (face-to-face) in the vice or with G-cramps, check that the lined-up edges and ends are square, then measure 13mm in from each end and *very lightly* knife-mark squared shoulder lines across the four face-edges. The pieces are then released and separated into pairs: two for dovetailing, which can be marked with a (D) and two to receive dovetail-sockets and pins, which can be marked with a (P).

On the two (D) pieces, I reinforce the faint 'datum' edge-marks with more incisive knifing and transfer them squarely and, again, incisively across the inner face-sides and the rear-edges – but not the remaining unmarked face-sides. These last two shoulder lines must be marked with a sharp carpenter's pencil.

On the two (P) pieces, the very lightly-knifed 'datum' marks on the face-edges (being only datum lines that will not be shouldered) must remain without deeper knifing, but they are used to transfer deeper-knifed, square shoulder-lines across the inner face-sides and – as on the (D) pieces – pencilled shoulder lines across the face-sides marked with a sharp carpenter's pencil.

Next, on the face of *one (D) piece only*, mark a pencilled centre line at each end across the width – between the shoulder and the end – equal to half the timber's thickness (6mm) up from the two shoulder lines. This is for setting out the dovetails with dividers.

## End-grain planing

*Figure 3.32*: The initial end-grain planing – mentioned previously – has to be done at this stage to create smooth, even (and square) surfaces for the seating of the dovetail template and the try square; and to give a clearer definition of the setting out marks made on the end grain. Even so, sometimes, especially on dark-coloured hardwoods, white chalk is rubbed into the end grain to highlight the pencil- or knife-marks. The planing (shooting) of the end grain can be done with a block plane or a finely-set smoothing plane (my preferred method), but a developed skill is needed to avoid *spelching* (breaking away the end

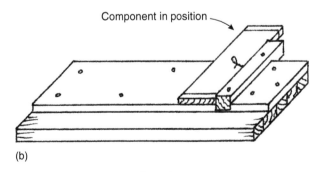

Figure 3.32 A shooting board.

fibres) and in retaining squareness. Alternatively, if cut with a fine saw, the ends can be sanded – or, as illustrated, they can be *shot* square on a traditional-type *shooting board*. Although not shown, a smoothing- or jack-plane, lying on edge and held from above, is slid backwards and forwards along the recessed edge of the shooting board whilst the component is held firmly with the back-palm of the hand against the right-angled stop.

## Production technique for four sets of tails

There are variations of production techniques, but I take only one of the (D) pieces – mentioned above – and set out two sets of dovetails (one at each end) on the face-side with my dividers' method. When marking the dovetail slopes, I continue the pencil lines down past the pencilled shoulders for at least 12mm, as this helps to visually align the saw to the angle within the otherwise shallow depth of cut.

The setting out of tails can also be done on the other (D) piece, if you prefer, but, professionally, craftspeople usually mark only one or two sets in a four-set production and use these as templates to cut the other two or three sets.

The dovetails are now nearly ready to be cut, but first take the other (D) piece and place it face-side to back-side and face-edge to face-edge against the marked-out template piece. Line them up lengthways and sideways, with particular attention to matching up the edge-marked shoulder lines, and bind them very tightly together with a band of masking tape near each end. You cannot rely wholly on this, so you will still need to check all edges for alignment as you place the two pieces in the vice, ready for dovetailing.

Next, square the tail ends across the end-grain of both (D) pieces and mark with a chisel-knifed carpenter's pencil. Then turn the pieces upside down and mark the other ends. As previously mentioned, if your pencil marks do not stand out too clearly on a particular species of wood, rub a stick of light-coloured chalk across the end grain before using the pencil.

## Cutting the tails

Now reposition the taped pieces vertically in the vice, with about 75mm upstand, and cut the tail-slopes carefully down to the shoulder lines. When quite close to the lines, move your head over the work occasionally whilst sawing and check the depth of cut on the rear face of the second (D) piece. If you are right-handed, cut all the right-hand slopes first, then all the left-hand slopes. This continuity of stance and repetition is believed by me to promote greater accuracy and speed. However, some woodworkers, who prefer to saw vertically, tilt the work in the vice one way and then the other. Others also use dovetail saws, but I prefer to use a gent's saw.

When both sets of tails have been sawn this way, release the work from the vice and remove the masking tape. Then apply the try square to the four pencilled shoulder lines on the face-sides (but with particular reference to the edge-knifed shoulder lines on the face-edges) and deeply knife all intermittent parts of the shoulder lines between the dovetails, i.e., the *pin sockets*, where the pins and half pins will be seated. These hit-and-miss knife cuts might easily overrun, but are controlled by the point of the marking knife starting in the kerf at the base of one saw cut and ending in the kerf at the base of another.

Note that some craftspeople knife the shoulder lines right across the face-side, not intermittently. Such changes in technique are just another interesting variation. I did it this way myself some years ago (perhaps when I was less pedantic) and evidence of this practise can often be seen on the lapped-dovetailed sides of antique drawers.

## Removing pin-socket waste

*Figure 3.33*: Back in the vice again, but this time with individual pieces, the waste-wood in the pin sockets (between the tails) is removed. As illustrated, this is usually done with a coping saw. First, though, a vertical cut is made roughly in the middle waste-area with a gent's saw to within 2 to 3mm of the shoulder line. The coping saw is used in this cut only (to limit the risk of damaging the tail cuts) and it is turned on a small arc at the bottom towards the tail, to remove half of the waste. It is then turned the other way to remove the other half.

The waste on the outer edges of the tails is removed with the gent's saw, carefully leaving about 0.5mm to 1mm for final chisel-paring down to the knifed shoulder lines. In softwood, these outer shoulders that seat the half pins are sometimes finished by saw only.

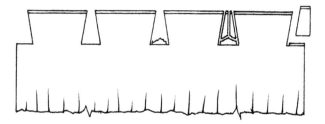

**Figure 3.33** Removing the waste material from the pin-sockets.

But, even with good sawing skills, this can lessen the quality of the joint.

The chisel-paring of these outer shoulders and of the remaining waste between the tails, in the pin sockets, is often reckoned to be done from both sides of the work, with the work lying flat on a chopping board of waste timber. But my experience has taught me otherwise. I find that if the end-grain fibres are pared and thereby pressured from one direction, then the other, they often break off in the middle area and lessen the integrity of the joint. Therefore, my method is as follows:

## End-grain chisel-paring technique

Whether making a dovetailed item in hardwood or softwood, I use an offcut of close-grained, dense hardwood as an underlying buffer/chopping board to remove the pin-socket waste. The tailed-end rests on this, face up on the bench and the end-grain waste is removed by near-vertical paring, taken through to the chopping board and gradually adjusted to vertical-paring back to the knifed shoulder lines. For the reasons already given, I do not turn the work over and pare from the other side – but there is a need to keep lifting and turning the work briefly to check how close you are to the knifed shoulder-line on the under-face.

If competent at vertical paring, this method works well and is fast and efficient. It also eliminates any risk of damaging the face-side shoulders. I use newly sharpened, ordinary bevel-edged chisels for this work, but the shallow, square side-edges of these do not ideally suit the acute angles of the pin sockets and often leave small upstands of waste in the corners. However, it is quite easy to modify a small-sized bevel-edged chisel by grinding and/or stoning the shallow side-edges from 90° to about 80°. Otherwise, Japanese chisels, with an isosceles sectional shape, can be used.

## Marking out the dovetail sockets

The dovetail sockets and pins required on the two (P) pieces of the box can now be marked from each correspondingly numbered set of dovetails. This numbering, usually encircled with (1) (1) to (4) (4), is pencilled on the face-sides of each end of the box to avoid mixing up the matched sets. And each set is cut separately. To do this, place each (P) piece in the vice, with the joint number (and face mark) in front. Place a 70mm block of wood – or a 04½ smoothing plane on edge (my method) – on the bench, initially up against the workpiece and adjust the work to meet the level of the plane's height. Then push the plane back from the (P) piece to act as a rear bearer and lay a dovetailed piece

on top, with corresponding numbers and face marks visible. Check that the (D) piece is seated evenly on the plane and the (P) piece in the vice.

Whilst supported like this, keep some hand pressure on top and adjust the piece carefully to line up the shoulder lines and side-edges. Because the edges are not easy to line up, you should test for side alignment with a small straightedge (a piece of wood measuring, say, 150 × 20 × 20mm) and then mark the tail sockets with either the specially-sharpened carpenter's pencil or a marking knife.

Sometimes I reinforce these marks with the aid of the dovetail template. But if you do this, you must realize that there may be slight inaccuracies in some of the dovetail slopes you have cut – so, be prepared to follow the first marks, even if it means making slight adjustments to the true-seating of the template as you re-mark.

Next, square down the marks onto the face side, at least 12mm past the pencilled shoulder line – for visual guidance mentioned previously – then roughly shade in the waste areas to highlight which side of the line to cut. This, in my opinion, is an essential practice for beginner and expert, as experience is no defence against a momentary lapse in concentration.

## Cutting the dovetail sockets

*Figure 3.34*: With a 75mm or so projection in the vice, cut the tail sockets down vertically and carefully to the shoulder line with a gent's saw. When starting the cuts (in the waste area!), keep up close to the end-grain knife-marks (or knife-like pencil marks) and take care in leaving a fraction of the pencil lines showing on each side of the pins. When complete, release the work from the vice and apply a try square to the pencilled shoulder line on the face-side (but with particular reference to the faintly-knifed shoulder line on the face-edge) and deeply knife all intermittent parts of the shoulder lines between the pins, i.e., the *dovetail sockets*, where the dovetails will be seated.

Then, back in the vice to finish the sockets in a similar way to the pin sockets, i.e., make a gent's saw cut in the mid-waste area and use this to remove the

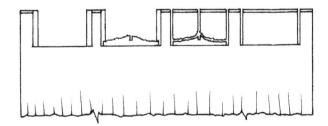

**Figure 3.34** Removing the waste material from the dovetail-sockets.

bulk of the waste with a coping saw. Then cut back to the deeply incised shoulder line with a wide bevel-edged chisel, by vertical paring onto a hardwood chopping board.

## Gluing up

Some joiner/cabinetmakers reckon that a set of hand-made dovetails, once finalized as separate components, should not be put together dry to test the fit, but should only be joined when being glued up. And whilst it is true to say that once you lock this type of joint together dry, there is a risk of damaging it and slackening the fit when you take it apart, I believe that – as with all joints – there should be a trial fit prior to gluing up. However, to eliminate the risk of shearing the short grain on the sides of the tails and/or spoiling the fit, it is essential to keep each component at right-angles to each other when the joint is being knocked apart with a hammer and a *hammering-block* (an offcut of waste-wood) and to avoid seesawing (creating an uneven, lateral separation) of the tails leaving the sockets.

The actual gluing up – if you have achieved good, snug joints – should be done with the aid of a hammer and a hammering-block. Cramping should be avoided. All squeezed-out glue should be wiped off with a damp cloth and, before setting aside, the box must be checked for squareness with a so-called *squaring stick*. This is a thin lath with an approx 30° speared end, which is used to check both diagonals of the box for equal length. By inserting the spear-point into the alternate inner corners and marking the other end of the stick against the other inner-corners, the squareness can be proved if both marks coincide. But if two separate marks are recorded, then the structure has to be gently distorted and retested until both diagonal lengths equal a central point between the first two marks.

## Lapped dovetails

*Figure 3.35*: There is no rigid rule regarding the size of the lap (or lip) covering the end-grain of the dove-tails. On drawer fronts (where such dovetails are most commonly used – and the fronts are made thicker to accommodate them), the lap is usually not more than a quarter of the front's thickness. However, it is often seen to be less, as craftsmen over the years seem to have regarded it as a measure of their skill to make the lap as thin as possible. Once the size of the lap has been decided, it is deducted from the drawer-front thickness to provide a measurement for setting up a cutting gauge. For example, say 20mm drawer-front thickness, minus 5mm lap, equals a gauge-setting of 15mm.

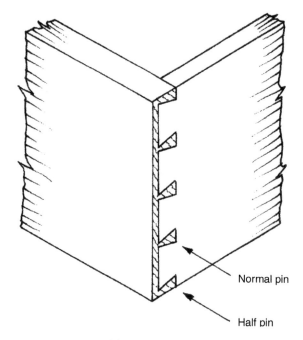

Figure 3.35 A set of four lapped-dovetails.

*Normal pin*

*Half pin*

## Setting out the tails and sockets

The stock of the 15mm-set cutting gauge is then used against the squarely-shot (planed) front-end of a drawer-side to prick the inner face from the end. This prick-mark is picked up with a marking knife and squared across the inner face. (Alternatively, the pricking and knifing can be avoided and the inner face can be marked across the grain with the cutting gauge). Either way, this shoulder line is then used to knife-mark both edges. In turn, these edge marks are squared across the face-side with a sharp pencil. A pencilled centre line is also established and the tails are set out and produced exactly as described for the through dovetails.

Setting up and marking the tails onto the vice-held, smoothing-plane-on-edge-height drawer front, is also exactly as described for the through dovetails. The only difference is that in addition to the tail-slopes being marked out, the precise ends of the tail sockets are also established with the pre-set cutting gauge from the *inner* face-side. (This being one of the rare occasions when gauging is not done from a face-side or face-edge).

## Cutting the lapped tail-sockets

*Figure 3.36*: Forming the *stopped* tail-sockets is started with a gent's saw (or dovetail saw, etc) by cutting the waste-side marked edges of the sockets carefully to two angles from the inner face, whilst the component

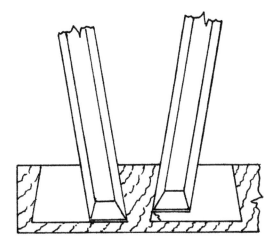

Figure 3.36 Corner-paring of stopped tail-sockets with left- and right-handed skew chisels.

is held vertically in the vice. The first angle (to the vertical plane) is the marked tail-slope, the second angle (to the horizontal plane) is the angle of cut required to come close to the shoulder line on the face and close to the cutting-gauge line on the end grain at the top. The waste area of each socket is then partly reduced with a coping saw, by making similar angled cuts to the horizontal plane from a mid-position one way and then the other, carefully up to the socket's saw-kerfs. To help visualize my description, when cut this way, the sockets will appear to contain rough-sawn chamfered edges.

In the next operation, the bulk of the tail-socket waste is chopped out by chisel and mallet across the grain on the inner face-side, then vertically with the grain, on the end, whilst the work is held in the vice. By this method, a hardwood *backing board* must be placed behind and the work must be positioned fairly low in the vice. I usually kneel down to do the horizontal chisel-and-mallet chopping, with one knee on a small offcut of carpet. But other positions and stances can be adopted. My technique is to start chopping across the grain in the mid-waste area of a socket, halfway up the rough chamfer, then I release the chopped fibres with a vertical cut to form a small inverted step. By alternate horizontal, then vertical chopping (paring) – 2mm to 3mm each time – the step is soon transformed to become a tail socket. This procedure is repeated in each socket and then the work is taken out of the vice to finalize the vertical paring (without the mallet) of each socket's end-grain up to the incised shoulder line. Ideally, at this stage, you need a left-hand and a right-hand *skew* chisel for paring out the acute-angled corners of the sockets. If you have a healthy-sized lap on the drawer-front, an ordinary 6mm or 10mm chisel will do the job without seriously damaging the lap – but, if you want to

compete with past masters in achieving very thin laps, skew chisels (as illustrated) can be made from ordinary bevel-edged chisels of the same width, with the cutting edges reground slightly to an approx 80° angle.

When the paring of the end grain in the sockets is nearing completion and you have reached the knifed shoulder line, hold the flat-sided point of the chisel firmly against the socket shoulders with finger-tip pressure against its extreme end, and hold a small try square behind to test the vertical-paring squareness.

The final paring (along the grain) and testing of the thin inner-lips of the lapped sockets is best done back in the vice. Again, hold the flat-sided point of the chisel firmly against the inside of the lap, with finger-tip pressure and hold a small (say, 150 × 20 × 20mm) straightedge on the face of the drawer-front and check that it is visually parallel to the chisel. If not, ease the lap socket accordingly.

## Gluing up a drawer

*Figures 3.37(a)(b)*: When finished, the sets of lapped dovetails – ideally – should only be partially tried together; not fully fitted or knocked together until they are glued up. And because of the cleaning-up allowance of 0.5mm to 1mm, after gluing, the final knocking together should be done with a wedge-shaped piece of hardwood and a hammer, working the thick end of the wedge in between the slightly-projecting pins. When doing this, the obtuse-angled side of the wedge should be kept close to the tail-sides to support the short grain.

If the lapped dovetails described above were for the front-sides of a drawer, then the back-sides would have

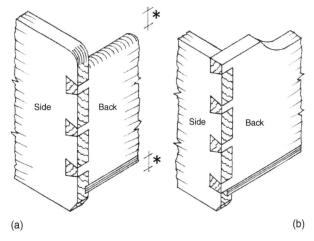

(a) (b)

Figure 3.37 (a) Top edge(*) reduced by at least 6mm to release the pump-effect of a snugly fitted closing drawer; and the bottom's height(*) should be not less than 6mm in hardwood, or 9mm in softwood; (b) Alternative top edge and tail-arrangement.

through dovetails; and the initial setting out of the tails would have to take into account the position of the plywood drawer-bottom of, say 6mm thickness. The grooves for this (the depths of which should be about ⅓rd of the side's thickness) would have to be accommodated within the drawer-front's bottom sockets and the sides' tails – to mask the groove's exit. And at the rear, it is common practise to reduce the depth of the drawer's back to allow the ply bottom to be slid into the front- and side-grooves before it is pinned or screwed (at closely-spaced centres of approx 75mm to 100mm) into the underside edge. It is also good practise to reduce the topside edge of the drawer's back to allow for a lowered edge of at least 9mm. This reduced edge is often rounded – and the two projecting, rear top-corners of the sides are either bevelled or rounded, as illustrated. The reason for the low top edge at the rear is to reduce the build-up of air pressure when the drawer is pushed in; as often this can cause other drawers or doors in a unit to be forced open.

Having pre-cut the plywood bottom squarely to size and made pilot holes in the rear edge for screwing, immediately after the drawer has been glued up I insert the (unglued!) snugly-fitted plywood bottom and – whilst keeping it pushed tight up into the drawer-front's groove – screw down the back edge. By this self-squaring method, no diagonal checks are necessary and the drawer can be set aside to allow the glue to set before cleaning up is undertaken.

## Edge fillets, drawer slips and muntins

*Figures 3.38(a)(b)(c)(d)*: Before leaving drawer-making, there are a few more details illustrated below which should be explained: (a) This shows a common groove in a drawer-side with a plywood bottom; the plywood is usually either 3.5mm- or 6mm-thick, according to the drawer's size; (b) This shows the addition of *edge fillets* which are top- and side-glued to the sides and front to give extra support to the grooved bearing; they also provide a wider running/sliding edge to the drawer-sides. Alternatively, short edge fillets of about 50 to 75mm length, can be used to give intermediate support and a glued connection to the ply bearing; (c) This shows traditional quadrant-shaped *drawer slips* which are glued and pinned to the inner drawer-sides. These were often used when it was felt that the drawer-sides were too thin to take a groove; they also provided a wider running/sliding edge to the drawer-sides; (d) This shows a traditional *muntin* which was used across the central, narrow span of the drawer's bottom to give support to the front and back of a very wide drawer. The muntin was lap-dovetailed to the under-edge of the drawer's front and rebated and screwed to the under-edge of the back. Drawer slips and muntins were

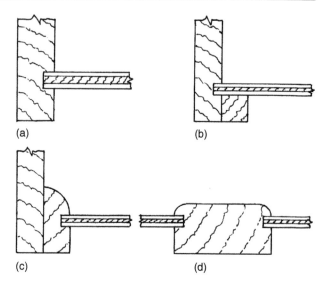

**Figure 3.38 (a)** Drawer sides (and front) grooved ⅓rd to receive plywood bottom; **(b) Edge-fillets** added to sides (and front); **(c)** Traditional quadrant-shaped drawer slip glued and pinned to drawer-sides; **(d)** Traditional muntin used on wide drawers.

more common when drawer-bottoms of solid timber were used. Such bottoms were usually 6 to 9mm thick and were often splayed off like raised-and-fielded panels on the undersides to fit the grooves. The grain ran from side to side and the front long-grained edge was glued into the groove; the opposite edge (slightly protruding) was open-slot-screwed to the underside, *unglued* back-edge to allow for shrinkage.

## Machine-made dovetails

*Figure 3.39*: In industry, with the aid of metal templates (shaped like sets of uniform, protruding fingers) and jigs, dovetails are mostly – if not entirely – made by machines nowadays. With dovetail-shaped cutters fitted to fixed machines or portable powered routers, the two components of a set of dovetails are clamped together in an offset position (equal to the width of one tail) and the tails and pins are formed in one machining operation. Because of this, they are equal to each other; i.e. of uniform shape and size. Also, this process produces semi-circular shaped inner-edged tails and inner-edged dovetail sockets. If the material being jointed is relatively thin, say 10mm, the semi-circular ends of the tails will be exposed, as illustrated in Figure 3.39. But if the material is thicker, the ends will be covered and lapped dovetails will be produced (similar in appearance to those illustrated in Figure 3.35).

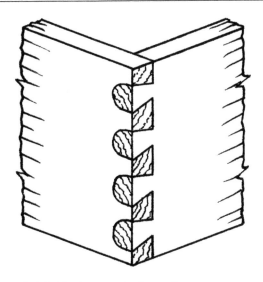

Figure 3.39 Machine-made dovetails.

## Bridle joints

*Figures 3.40(a)(b)*: These joints in their original, three-layered form as illustrated, are traditionally known as *bridle joints* – sometimes referred to as *open-ended mortise-and-tenon joints* – but in their five-layered form (or more), these framing joints are referred to as *comb-* or *finger-joints*. The multi-layered comb joint has a decorative appeal in presentday cabinetmaking and is still used for jointing the framing members of modern-style casement windows (as described in Chapter 4). These joints are ideally made by a machining process, but they are also quite easily made by hand. The latter being a skill-challenge in ripping accurately along the grain on the waste-side of the multiple gauge lines.

## Dowelled frame joints

*Figure 3.41*: Dowelled joints are often used as a substitute for mortise and tenon joints, but are generally regarded as being inferior to them. And although I am inclined to agree with this, I do so not because of the joints failing structurally (this should not happen if adequately-sized dowels are used; i.e. the diameter should be approx two-fifths of the frame's thickness), but mainly because I have occasionally found these joints failing via the glue being skimped in their manufacture. Glue-skimping being more critical to dowelled joints than to (wedged) mortise and tenon joints.

Traditionally, these joints were always regarded as being difficult to achieve by hand – and there were two reasons for this. One being that the drill bit – whatever type you use – tends to be taken slightly away from its marked position when up against the hard and soft fibres of the timber. The other reason is that even with another person sighting the brace or drill and advising you of its verticality (in two vertical planes!), drilling precisely-positioned holes aligned to the face side *and* at right angles to the edge, is virtually impossible.

However, apart from modern, fixed machining-processes, nowadays there is a variety of inexpensive, patented dowelling jigs that can be bought. These are clamped onto the work and the jig's different sized steel bushes keep the lip-and-spur bits in position and at right-angles to the joint's surfaces. Note that if a frame, such as a small door, requires its inner edges to be rebated, grooved or moulded, the holes for the dowelled joints must be drilled prior to the edge treatment.

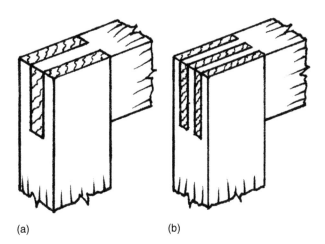

(a)  (b)

**Figure 3.40 (a)** Bridle joint **(b)** Comb- or finger-joint.

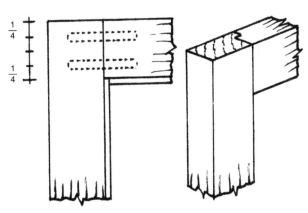

**Figure 3.41** Dowelled joint to top rail of rebated door; note that the dowels' depth. should be a half or two-thirds the width of the stiles.

# MISCELLANEOUS FRAME JOINTS

## Housing joints

*Figures 3.42(a)(b)(c)(d)*: The four types of housing joint shown here are (a) *Through housing*; (b) *Tongued (or shouldered) housing*; (c) *Stopped housing*; and (d) *Dovetailed housings*. Although there are numerous uses for these joints in joinery and cabinetmaking, (a) and (b) are commonly used for jointing the heads of door-linings; (c) is used to house the treads and riser-boards into stair-strings – and to act as neat and inconspicuous *bearing recesses* for shelves; and (d) has a traditional use in cabinet-making for jointing the ends of drawer-sides that are inset and emanate from an oversailing drawer-front. As illustrated, such dovetailed housings may have a single- or a double-splay.

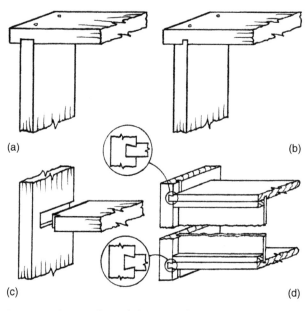

(a)   (b)   (c)   (d)

Figure 3.42 **(a)** Through housing; **(b)** Tongued (or shouldered) housing; **(c)** Stopped housing (shown separated); **(d)** Stopped dovetailed-housings (single-sided and double-sided) shown here on drawer-sides.

## Halving or half-lap joints

*Figures 3.43(a)(b)(c)(d)*: The four types of joint shown here are (a) *90° corner halving (or corner half-lap)*; (b) *T-halving (or T half-lap)*; (c) *Dovetailed halving (or dovetailed half-lap)*; and (d) *Mitred halving (or mitred half-lap)*. These traditional joints have a non-specific use in joinery and joints (a) and (b) are more commonly used in first-fixing carpentry work to join timber wall-plates, etc, together.

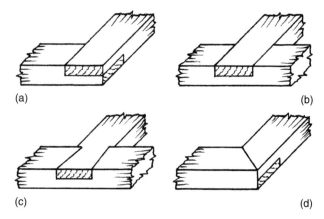

(a)   (b)   (c)   (d)

Figure 3.43 **(a)** 90° corner halving (or corner half-lap); **(b)** T-halving (or T half-lap); **(c)** Dovetailed halving (or dovetailed half-lap); **(d)** Mitred halving (or mitred half-lap).

## Hand-skills techniques for housing- and halving-joints

*Figures 3.44(a)(b)(c)(d)(e)*: With all framing joints, there is a decision to be made whether to mark out the cross-grain shoulders with a sharp pencil or a marking knife on the face-sides and edges. To some extent this depends on the type of joinery being made and the skill of the woodworker, regardless of whether softwood or hardwood is being used. With skilful tenon-saw work, shoulders and sides of housing- and halving-joints, etc, can be cut accurately against a sharp pencil-line. However, if you want to be certain of a good fit and have perfected – or want to practise – the skill of *vertical paring*, excellent shoulders can be achieved by marking deeply with a sharp marking-knife instead of a pencil, then sawing and leaving about 1mm (or less) of waste to be pared off with a bevel-edged paring chisel, as illustrated at 3.44(b) and (c) below.

As illustrated at (a), after making three or more cross-grain tenon-saw cuts, vertical paring can also be practised when removing the waste wood from halving joints. This illustration indicates slightly-angled paring from each side onto a waste-wood chopping board – the angled paring being necessary to avoid *spelching out* (breaking away) the fibres on the opposite edge. After gradual reduction of 2 to 3mm per paring, down to the gauge line on each side, the middle peaked area is then vertically pared to create a flat surface; this is tested with the edge of the chisel, used as a straightedge.

In most textbooks, this paring technique is usually shown with the work held in a vice – with the chisel at about 15° to the horizontal – and being turned around to reduce the waste on each side of the initial peaked shape, before being horizontally pared to create the

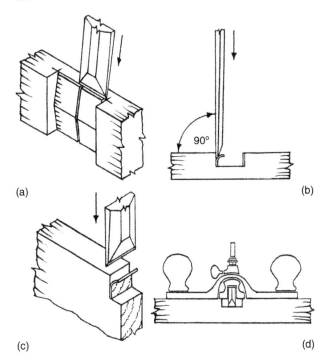

(a)

(b)

(c)

(d)

**Figure 3.44 (a)** Angled paring of a half-lap joint onto a chopping board; **(b)** Vertical paring to sides of a housing; **(c)** Vertical paring to recess of stopped housing; **(d)** Hand-routed housing.

flat surface. Providing the rule is practised to *always keep one's hands (and body-parts) behind cutting-edges*, both methods are safe, but the text-book method is more suitable for inexperienced woodworkers. My method of vertical-paring is quicker, but – because the work is not in a vice and is being held onto the chopping board with the palm-edge of one hand whilst chiselling (at an angle) with the other – if not held firmly enough, the work can tilt over whilst chiselling.

With reference to Figure 3.42(c), the ends of shelves are usually housed-in to the shelf-unit's cheeks by one-third the cheek's thickness – but the appearance of housing grooves in the face-edges is traditionally regarded as being amateurish, so they are *stopped* away from the edge by an optional amount which should not exceed the shelf's thickness. This results in a small step having to be removed from the front ends of the shelves. A typical housing-step is shown in Figure 3.44(c), with its knife-marked, sawn shoulder being vertically pared. *Through housings* and *stopped housings* shown in Figures 3.42(a) to (d), can all be cross-cut with a tenon saw – to a finish, or with an allowance for paring to a more-precise finish – then horizontally chisel-pared across the grain to within a few millimetres of the gauge lines and finished off with a Stanley-type Router No.71 hand plane, illustrated in Figure 3.44(d), with its extended cutter in a grooved housing. Note that *stopped housings* formed

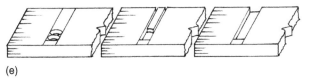

(e)

**Figure 3.44 (e)** Left-to-right sequence of 1) recessed drilling of approx housing-depth; 2) Shallow holes chisel-pared to form recess, then saw cuts made; 3) Waste area chisel-chopped and routered to a finished stopped-housing.

by hand, have to be partly formed with a *sunken recess* at the stopped end to allow the toe of a tenon saw to run freely when cutting the sides of the housing. Once the width of stopped housings is pencil- or knife-marked across the cheeks and the gauge marks for the housing-depth and stopped-ends are done, the sunken recesses are formed up to the face-side gauge lines by boring shallow holes with a Forstner-type bit, then squaring them up by chisel-paring (as shown in Figure 3.44(e).

The dovetailed ends for the stopped dovetailed housings shown in Figure 3.42(d) can be formed by cutting the shallow shoulders with a gents' saw or a cutting gauge, then by using a long bevel-edged paring chisel or a shoulder plane to form the sloping sides of the dovetails.

If wishing to reduce some of the hand-skills' work on the housing- and halving-joints mentioned above, the half-lap joints could be *deeped* on a band saw and the housings could be trenched out with a portable powered router.

## EDGE JOINTS

*Figures 3.45(a)(b)(c)*: Boards that are not wide enough for a particular job have to be edge-jointed and there are various ways of doing this. However, before dealing with these, it must be understood that the wider the board is in its original width, the more likelihood there is of it being adversely affected by shrinkage.

Depending on the equilibrium moisture-content of the timber, there will be some shrinkage in width and a lesser amount in thickness. Also, depending upon

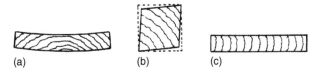

(a)

(b)

(c)

**Figure 3.45 (a)** Cupping **(b)** Diamonding **(c)** Quarter- or rift-sawn board.

how the timber has been cut (converted) from the log, wide boards may also take on a shape known in the trade as *cupping*. This, as illustrated in Figure 3.45(a), is when the board becomes convex on the heart side.

By examining the end grain of a piece of timber, you can usually detect what is likely to happen by remembering that the greatest amount of timber-shrinkage takes place in the direction of the annual growth rings. Therefore, the longest annual rings on a particular cross section of timber will shrink more than the shorter rings on the same section. As illustrated in Figure 3.45(a), this will alter the shape of the timber. In this example, the board has been cut *tangentially* and, because of this, it is also likely to shrink more in thickness on the outer edges than in the middle area. This is because timber shrinks more in the sapwood area than in the middle heartwood.

Figure 3.45(b) shows the effect of shrinkage known as *diamonding*, which can happen when the annual rings run diagonally across a section of timber. Again, the greater shrinkage along the longest annual rings has caused this.

Figure 3.45(c) shows a section of timber which has been *quarter sawn* or *rift sawn*. This type of log-conversion creates short annual rings of equal length, radially square to the face, which will ensure equal and minimal shrinkage without distortion.

## Butt joints

*Figure 3.46*: Butt joints are sometimes referred to as *rubbed joints*, as this was the traditional method used for joining two boards edge-to-edge with reheated Scotch glue, a toffee-like glue made from recycled animal bones, sinews and hides, etc. The glue was applied to both the prepared edges; then, with one board held in the vice, the other placed on top, the top board was rubbed backwards and forwards a few times until the excess glue was exuded and the boards were stuck firmly together.

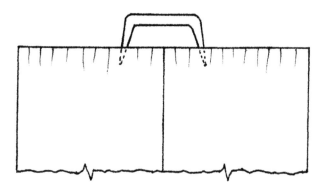

Figure 3.46 'Iron dog' across butt joint (at each end).

Sometimes, especially if there was a shortage of available sash cramps in the workshop, a small *iron dog* would be driven across the joint at each end, into the end grain (as illustrated), to help hold the joint together until the glue had set.

## Gluing with PVA glue

When the edges have been hand-planed and there are no rotary cutter-marks from planing machines, this traditional method of jointing has been scientifically proven to be stronger than machined-and-glued butt jointed edges. Although animal glues are not used for jointing nowadays, rubbed butt joints can still be successfully made by using PVA (polyvinyl acetate) adhesives. There is not the initial 'grab' experienced with animal glues, but there is enough – provided only one of the jointed edges is glued.

Of course, holding the freshly-jointed boards together with a few sash cramps would be better than relying wholly on the iron dogs and would also promote alignment across their face-sides. However, the top edges of the metal cramps can stain the wood badly – where the glued-surfaces are still wet from being wiped with a damp cloth and/or more glue has exuded – so cover the cramps' metal edges with masking tape or pieces of paper before laying the work into them.

This method requires only the face-side and face-edge of each board to be planed true prior to gluing. When planing the edges, both boards can be held together in the vice, face-to-face, and separated for testing. The main point is that each board must have a perfectly straight edge. Being square to the face-side is ideal, but, providing the face-sides are together when planing (or both on the outer surfaces, if the boards have been thicknessed), a slight lean one way or the other will still achieve perfect face-alignment across the boards when they are put together. Testing for straightness can be done by eye, with a straightedge and/or by trying the edges together. Planing should be done with a long plane – preferably a *try plane*, but a *jack plane* will do the job. Unless very short lengths of board are being shot for jointing, a *smoothing plane* is not recommended.

## Gluing technique

When the boards' edges are together and judged to be ready for gluing, the mid-area of the face-side should be freehand-marked across the joint with two closely-spaced pencil lines. One piece is then placed in the vice and glue applied; the other (dry) piece is placed on top and rubbed backwards and forwards (by about 75mm) a few times until a resistance to movement is felt. At this stage, push slowly and stop when the two mid-area marks are lined up. Whilst rubbing, your fingertips should protrude, if possible, below the thumb-held holding of the top board, to act as fences guiding the two faces to a flush alignment.

After carefully wiping off the exuded glue with a damp linen cloth, the jointed boards should be left for about five minutes to achieve an initial set. Then they should be released from the vice and very carefully laid and cramped up in two, three or more sets of cramps, according to the boards' length. Prior to this, as mentioned above, the top, mid-area edges of the cramps should be covered with masking tape or paper in the area of the wet joint.

After being left overnight, or for at least 12 hours to set (although bear in mind that the manufacturers of PVA adhesives usually recommend 24 hours setting to achieve full strength), the jointed boards should be ready for minimal resurfacing prior to being gauged and planed to width, then thicknessed to the finished size.

## Gluelam joints

*Figure 3.47*: Gluelam (glued and laminated) construction (also referred to nowadays as 'engineered wood') uses any number of narrow strips of timber (laminae) – of softwood or hardwood – with their side edges glued and laminated together to build up whatever width is required. Providing the appearance of the multiple edge-joints is acceptable on a particular job, and the growth rings of each piece have been placed to oppose each other, as illustrated below (to combat unequal shrinkage and surface distortion), this is an excellent way of building up and stabilizing wide surfaces.

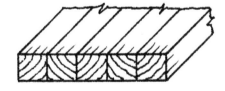

Figure 3.47 Gluelam board or 'engineered wood'.

## Origins of gluelam

This method of edge jointing developed in the construction industry (predominantly in countries other than the UK) about five decades ago and was used by specialist manufacturers of gluelam beams and roof trusses in the form of arched ribs. The laminated arch shapes were formed in large jig arrangements, using

synthetic resin adhesives. By virtue of being bonded together, the relatively thin laminae held their curved shape after the adhesive had set.

This principle was also used by manufacturers for producing bentwood components in furniture making. For joiners and cabinetmakers interested in forming bentwood (curved) shapes, the thickness of the laminae is determined by dividing the radius by 150. For example, a bentwood segment of 600mm radius = $600 \div 150 = 4$mm thick laminae. When making the jig or formers, it must be remembered that the radius for the inner, (convex) former is different to the radius for the outer, (concave) former. The difference equals the overall laminated size of the component. This has to be added to the inner radius – or subtracted from the outer radius.

## Finger-jointed/engineered wood

Nowadays, gluelam construction is used widely by furniture makers and manufacturers of hardwood kitchen-worktops, etc. In straight (uncurved) work, the separate laminae are of relatively short lengths which are either *finger jointed* or butt jointed on their ends, before being butt jointed together on their sides. For obvious structural reasons, the end-joints are randomly staggered. The lamina are usually 18mm to 25mm thick across the face of the built-up board, but there are no hard and fast rules about this – except to say that the board is less prone to distortion with thinner laminae.

Gluelam work, whether for forming wide boards or making shaped work, is easily achievable by the individual craftsperson. The joints on the ends of the end-jointed lamina – if required on lengthy work – can be neat end-grain butt joints or neatly-cut 45° splay joints.

## Dowelled-edge joints

*Figures 3.48(a)(b)*: The preparation of the boards' edges for these joints follows the same initial procedure described above for butt joints. However, although dowelled-edge joints were traditionally used to reinforce the joined edges and create flushness on the face side, they are – as mentioned above for *dowelled-frame joints* – difficult to achieve by hand, but relatively easy with the aid of an inexpensive dowelling jig and/or with the availability of a pillar drill.

Before moving on to current methods of edge-jointing, it must be mentioned that un-dowelled, rubbed butt joints are still popular in small workshops, but they are awkward to achieve and handle beyond a certain length – of about a metre. This is where the dowelled edge can help: on jointing longer boards.

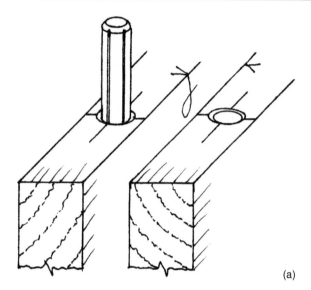

Figure 3.48 (a) Dowelled edge and opposite counter-bored edge.

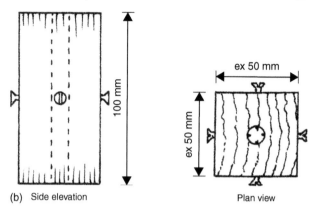

Figure 3.48 (b) Purpose-made, hardwood dowel-grooving block.

And for those who want to use dowelled edges for this reason – and have a dowelling jig and/or access to a pillar drill – the following notes apply.

Machine-serrated beech joint-dowels can be purchased, or you can make your own from readily available 1.8m lengths of hardwood or softwood dowel rod. These will need to be cut into 65 to 75mm lengths and their ends should be chamfered or rounded off. Then the dowels need to be grooved to allow trapped air and glue to travel up the dowels when the joint is being made.

## Dowel-grooving block

*Figure 3.48(b)*: One way of doing this is to make a *dowel-grooving block* from a short piece of hardwood with a hole drilled through its length to receive the dowels. Up to four screws are screwed into the block, so that they just break through the surface radially into the drilled dowel-hole, as illustrated. When the pre-cut dowels are hammered through the block, the sharp points of the screws score v-shaped grooves down their length. Hammering more than two dowels into the block, releases the first dowels that were entered.

When the boards to be jointed have been face-sided and edged, the dowel-positions should be set in from each end by about the length of the dowels and spaced apart by at least 225 to 300mm. These positions should be squared and marked across both edges whilst the boards are held together in the vice or with G-cramps; a short, central gauge line, made on each face edge, bisects the squared pencil lines.

After being carefully jig-drilled or pillar-drilled

(to a few millimetres over half of the dowel's length) the holes should be countersunk slightly to remove any burred fibres and to facilitate easy entry of the dowels (especially if they are slightly misaligned). To make the joint, place one board in the vice and squeeze a small amount of PVA adhesive into each hole. Tap in the dowels and wipe off the exuded glue. Place the second board in the vice and squeeze the glue along the edge and into the holes as before. Then spread the glue with a brush to cover the entire edge and bring the joint together whilst the board is still in the vice. Knock the joint together evenly with a claw hammer and a *hammering/cushioning block* before laying the jointed boards onto pre-positioned (masking-taped or paper-lined) sash cramps. Apply even pressure and then wipe off the excess, exuded glue and leave to set.

The diameters of dowel rod commercially available vary from 6mm to 25mm and are usually in 2.4m lengths. The diameter required for jointing a particular thickness of board is as stated for dowelled frame-joints, i.e. approximately two-fifths the boards' thickness. Example: 20mm board ÷ 5 = 4, 4 × 2 = 8. Therefore, 8 or 9mm diameter dowels needed for 20mm-thick boards.

## Biscuit jointing

*Figure 3.49*: Again, the initial procedures described above for butt joints and dowelled-edge joints also apply to biscuit jointing. The only difference, of course, is that instead of pre-positioned dowels, so-called *biscuits* are glued into snug-fitting, segmental-shaped grooves, made with a portable biscuit-jointing machine. Like dowels – but much speedier and potentially more accurate – they reinforce and align the boards being joined. Most biscuit jointers seem to have good safety features and are easy to set up and use.

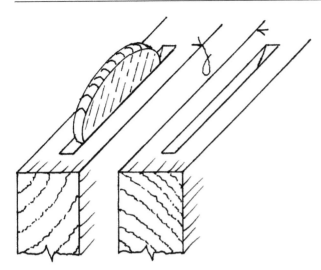

Figure 3.49 Open biscuit-jointed edges. Note that the elliptical-shaped biscuit is positioned in the centre of the segmental-shaped groove.

When manufactured, the elliptical-shaped biscuits are pressed and cut to shape from *feather-grained* beech and come in two sizes; feathered-grain is a traditional technique of arranging the grain at 45° across the face of the timber. This is done to eliminate the risk of the thin biscuits splitting along the grain when positioned midway within a joint. The grain direction is indicated in the illustration above.

## Loose-tongue jointing

This traditional tongue-and-groove method of reinforcing and aligning edge joints, again follows the basic procedures for preparing the edges. Then each edge is grooved throughout its length to receive a *loose* (i.e. separate) tongue and, when gluing the edges, this is glued into the grooves prior to cramping up.

The grooves can be cut on the following machines: 1) a circular saw-bench (if the saw's spindle can be lowered enough); 2) on a spindle moulding machine with a so-called *wobble-* or *drunken-saw*; 3) with a portable powered router – or 4) even on a biscuit jointer used as a mini portable-saw. An advantage with the last three methods mentioned, is that *stopped-grooves* can be made near each end if a concealed tongue is required for aesthetic reasons on a particular job. But if none of these fixed- or portable-machines is available, the grooves can be done by hand with a metal plough plane. For small jobs, I use either a Record 050 plough plane or a far superior Stanley 45 combination plane.

The main skill element in using plough planes is to do with keeping them in a constantly vertical position whilst planing. Starting at the front of the timber, the

plane is worked regressively backwards until an initial shallow groove is established. Then the end-grain appearance of the groove should be checked before proceeding, as this will reliably indicate whether you need to adjust the lean of the plane to achieve verticality. Although the fence of the plane is held tightly against the face side of the timber when the plane is being used – which should keep the groove upright – in reality the plane can develop a lean to one side or the other.

Traditionally, the loose tongues were made from selected timber to the thickness of the tongue required. This critical size was checked with a hand-made gauge called a *mullet* – which was a small offcut of timber (about 200mm long) with the required groove run in it to act as a gauge when slotted onto the tongue-material to test its thickness. The aim was to achieve a snug fit, not too loose and not too tight.

## Types of tongue

*Figures 3.50(a)(b)*: Once the snug fit was achieved, the material was cut to the width of the tongues required. Two types were used, as illustrated; one being *cross-grained*, whereby the grain ran at 90° across the tongue's width – at right angles to the joint line. The other being *feather-grained*, whereby the grain (as explained in biscuit jointing) ran at 45° across the tongue's width – at an acute angle to the joint line.

Both of these methods gave greater resistance to any side pressure applied to the made-up tongued joint, but the feather-grained – by being cut at 45° across the grain – produced longer lengths of tongue from relatively narrow, pre-thicknessed boards. Long-grained tongues, where the grain runs lengthwise – parallel to the joint line – were reckoned to be relatively less strong.

If the tongues being produced were not too thick – say not more than 6mm – they were cut by setting

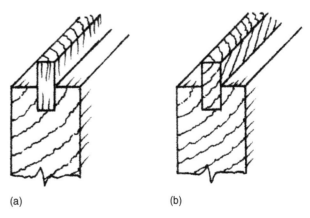

(a)                                             (b)

Figure 3.50 (a) Cross-grained tongue; (b) Feather-grained tongue.

up a cutting gauge to the tongue's required width and then by scoring deeply on the face side and the back side. I find that this method (depending on the density of the timber) needs additional, final cutting by running a sharp Stanley (craft) knife along the cut gauge lines.

## Plywood tongues

*Figures 3.50(c)(d)*: Another type of tongue which evolved from the two described above, is made by cutting off strips of plywood and removing the sharp arrises. Again, a cutting gauge can be used. As illustrated, the grain-direction of the plies needs to be considered, to allow the greater number of plies to be at 90° across the tongue's width – at right angles to the joint line. This relatively modern tongue also provides good resistance to any side pressure applied to the made-up tongued joint and is less time-consuming to make. The only downside is in making the groove to suit the tongue, which can be more problematic than the other way round.

There are usually only three thicknesses of plywood used for tongues, which relate to the thickness of board being jointed. These are 3.5mm (⅛"), 6mm (¼") and 9mm (⅜"). The most commonly used – to my knowledge – is the 6mm. Although there are no hard and fast rules regarding the size of the tongue and whether one or two should be used, the thickness of the tongue for a single-tongued joint should approximately equal a quarter of the board's thickness. For example, a 16mm-thick board ÷ 4 = 4mm; this would approximate to using a 3.5mm plywood tongue; a 20mm-thick board ÷ 4 = 5mm; this would approximate to using a 6mm tongue. Using the rule the other way round, a 9mm tongue × 4 = 36mm. Therefore, 36mm (or the nearest commercial finished size of 34mm) is the approximate maximum thickness for 9mm tongues. Beyond this – unless you use

thicker plywood-tongues – I believe that two tongues are required. Thicknesses of board, therefore, in excess of 36mm, should have two tongues equal to a quarter of the board's thickness, with the remaining thickness divided by three into the mid- and outer-area shoulders. For example, a 45mm-thick board ÷ 4 = 11.25mm (for two tongues) divided by 2 = 5.62mm per tongue, which approximates to 6mm plywood tongues. The remaining thickness = 45 − 2 × 6 = 33 divided by 3 = 11mm for each outer area and the mid-area shoulders. Although – again – there are no hard and fast rules regarding the width (or depth) of tongues, they should not be too deep – or too shallow. As a guide, once you have worked out the thickness and position of the tongue(s) in relation to the board's edge-thickness, the half-depth of the tongue should approximately equal the width of an outer shoulder. In the example above, this was 11mm. So, 11 × 2 = 22mm tongue.

## Tongue and grooved edge joints

*Figure 3.51:* The tongue of this traditional edge-joint, as illustrated, is formed from the material itself. Although such joints can be made with hand planes or portable powered routers, etc, because of the work involved in producing the tongued edge and the likelihood of uneven shoulders, they are usually regarded as a job for a *four-side moulding/cutting machine* – usually referred to as a *four-cutter machine*.

## 'F' Joints

*Figure 3.52:* This type of edge joint is a modern version of a tongue-and-groove joint, designed to suit a machining process. As illustrated, the joint resembles the letter 'F', which is a good interlocking configuration and also increases the glue-line for greater strength. Furthermore, it is not unattractive in its end-grain appearance. It would not be impossible to make a joint like this by hand, but it would be exacting, tedious and time-consuming. Realistically,

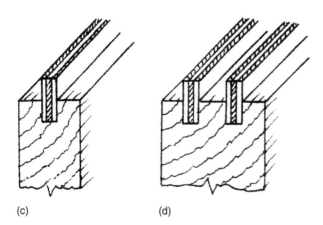

(c)  (d)

**Figure 3.50 (c)** Single plywood tongue; **(d)** Double plywood tongue.

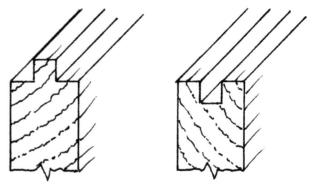

**Figure 3.51** Traditional tongue-and-grooved edges.

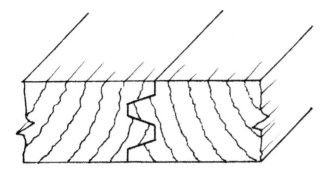

Figure 3.52 Machined 'F' joint.

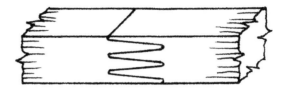

Figure 3.53 End-grain finger-jointed lamina.

the shape of the F would have to be modified to cut out the slopes (which would destroy the design feature that facilitates easy assembly).

# END JOINTS

There are two main reasons for end-jointing timber. The first obvious one being to make it longer, beyond its commercially-available length for a particular job. The second reason is to join certain joinery components together with end-to-end joints. Such joints are used on shaped heads of doors and doorframes, windows and window frames, etc, and on so-called *continuous handrailing* used on geometrical stairs. All of these items listed under the second reason will be covered here separately in the relevant chapters. End joints for the first reason are given below.

## Finger-jointed laminae

*Figure 3.53*: Manufactured gluelam or 'engineered wood', which uses a number of side- and end-butted (or finger-jointed) narrow strips (laminae) of softwood or hardwood, is not only useful for widening boards – as mentioned under Edge-Jointing above – but is also very useful for lengthening boards. Any lengths of laminae can be finger-jointed, as illustrated above, and built up with the joints randomly staggered to produce whatever widths and lengths of board required. This technique of end-jointing is the only modern means of producing lengthened boards that do not rely on other means of support.

## Supported end joints

*Figures 3.54(a)(b)(c)*: Other relatively modern, end-lengthening joints, such as for counters, bar tops and kitchen worktops, which need the support of the substructure, are usually bonded together with a waterproof adhesive or silicone sealant, brought to flushness with two or three biscuit joints and pulled together with two or more panel-bolt connectors on the underside.

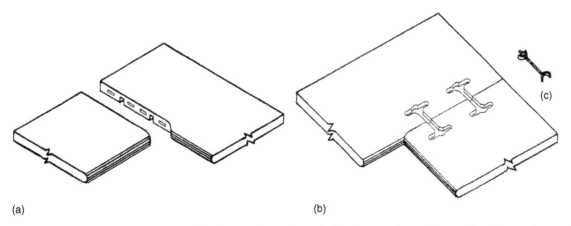

(a)  (b)  (c)

Figure 3.54 (a) Separated, routered end-joint of a right-angled kitchen worktop, showing four biscuit slots and two panel-bolt slots; (b) Joined-up underside of the worktop, showing the 'T' or bone-shaped recesses routered out to take the panel-bolt connectors – one of which is shown at (c).

# 4

# Making traditional and modern windows

## INTRODUCTION

Although uPVC (unplasticized polyvinyl chloride) windows and exterior doors with double-glazed sealed units have proliferated in the UK in recent decades, it seems likely that this trend may change soon. Research informs me that many window designers, seeking to cut carbon emissions and improve overall thermal performance are switching back to timber as a renewable and sustainable material with natural insulating properties – and are using low-E (emissions) glass in the sealed units. This renewed interest in timber windows has arisen from the Government's drive to reduce the carbon-footprint of houses by 2016; and a need to meet the recently amended Building Regulations Part L1A and L1B: *Conservation of fuel and power in new and existing dwellings*.

It will not be easy to overcome the tarnished image that many people have of timber windows (usually gained from expecting them to survive maintenance periods far in excess of the required 3 to 4 years). However, many manufacturers are now upgrading their designs. This includes using laminated timber sections for stability and extra strength and using quality redwood (such as Douglas fir) for durability and longevity. Timber is also treated with water-based preservatives that usually contain biodegradable fungicides and insecticides; and the completed windows can be given high-performance base- and top-coats of water-based paint or stain before leaving the manufacturer. Note that water-based coatings are now believed to give more protection than solvent-based coatings for exterior woodwork. However, where greater durability is required, a powder-coated aluminium cladding can be applied to the external face of the windows – especially in places where periodic repainting is impractical, i.e. medium- and high-rise buildings.

Whether the legal demands of the amended Building Regulations' Approved Document Part L1B will affect the work of the individual small-works joiner/installer of replacement windows is not yet known, but if not, my recent research conveyed that there is still a fairly big demand for the replacement (renewal) of like-for-like (upgraded) wooden windows – especially so-called boxframe windows with top- and bottom-hung sliding sashes. Wooden, stormproof casement-windows, with improved air- and water-tightness are also being used in newly-built houses. The upgrading of windows will be covered at the end of the chapter.

## BOXFRAME WINDOWS AND DOUBLE-HUNG SASHES

Boxframes in their original form were designed to accommodate and conceal the cast-iron or lead *sash weights*, which counterbalance each glazed sash window. The balanced sashes (which allow and retain a variety of open- or closed-positions) are moved up or down in two side-channels separated by *parting beads*. The sashes are linked to the sash weights by means of *sash-cord* or (for large, heavy sashes) *sash-chain* passing through two pairs of *sash pulley-wheels* sunk into the upper area of the *pulley stiles*. A removable *pocket* is cut into the base of the pulley stile on each side, concealed behind the side-edges of the bottom sash, to allow access to the weights for initial hanging and renewal of broken sash-cords. Note that another type of sliding-sash window (covered later) uses *spiral balances* instead of balanced sash weights.

### Details of a boxframe and sashes

*Figure 4.1(a)*: There are many component parts to these windows and the elevational drawing at Figure 4.1(a) indicates the various detailed sectional views that should be studied and referenced to the list of *common parts and sizes* given under that heading.

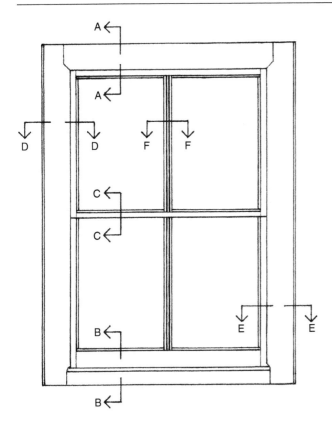

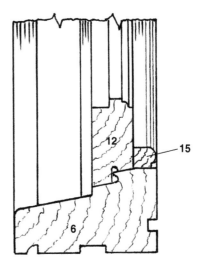

Figure 4.1 (c) *Vertical section B-B* showing (6) the grooved, rebated, weathered and anti-capillary throated sill; (12) the bottom-rebated and anti-capillary throated bottom-sash rail; and (15) staff bead. Note that, because of modern silicones and sealants, the underside grooving of boxframe-sills can be omitted nowadays.

Figure 4.1 (a) *Exterior elevation* of a boxframe window, with section-lines indicating the separate, detailed views.

## Common parts and sizes

*Figures 4.1(b)(c)(d)(e)(f)(g):* The vertical- and horizontal-section views at Figures 4.1(b) to (g) below,

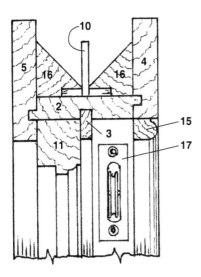

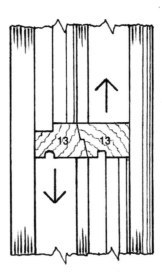

Figure 4.1 (d) *Vertical section C-C* showing (13) the rebated-splay meeting rails. Note that the main reason for the rebate was to stop intruders opening the sash fastener(s) with a thin-bladed knife-like instrument.

Figure 4.1 (b) *Vertical section A-A* showing (2) pulley-stile head; (3) square-edged draught bead; (4) inner-lining head; (5) outer-lining head; (10) wagtail; (11) top sash rail; (15) staff bead; (16) glue blocks; and (17) sash pulley wheel.

through the boxframe window and sashes at 4.1(a) above, are numbered to relate to the separate parts in the following list:

1.  *Pulley stiles*, 94 × 20mm par for 34mm-thick sashes (or 114 × 20mm, if thicker sashes of 44mm par are used);
2.  *Pulley-stile head*, same sizes as above – but minus

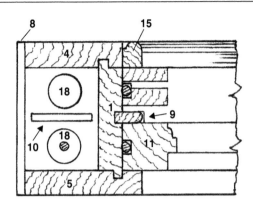

Figure 4.1 (e) *Horizontal section D-D showing* (1) pulley stile; (4) inner lining; (5) outer lining; (8) back lining; (9) parting bead; (10) wagtail; (11) sash stile of top sash (with stopped-groove for sash cord); (15) staff bead; and (18) high and low sash-weights. Note that the bottom sash (as seen on the top of its meeting rail, between the parting- and staff-bead) shows the end-grain detail of a dovetailed joint instead of the appearance of a more common joggle; this is explained further on in this chapter.

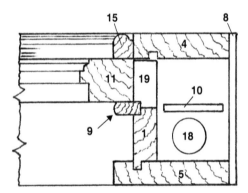

Figure 4.1 (f) *Horizontal section E-E showing* (1) pulley stile; (4) inner lining; (5) outer lining; (8) back lining; (9) parting bead; (10) wagtail; (11) sash stile of bottom sash (note that the stopped sash-cord groove does not run down this far); (15) staff bead; (18) sash-weight to the top sash; and (19) the access pocket – note that the technique for forming this, is detailed further on in this chapter.

the parting-bead groove, unless grooved for a *draught bead*;

3.  *Draught bead*, 21 × 9 or 10mm par (similar to a parting bead, but usually without the rounded front-edge;

4.  *Inner linings* (including the *inner-lining head*), 80 × 20mm par;

5.  *Outer linings* (including the *outer-lining head*), 96 × 20mm par;

6.  *Sill*, 120 × 70mm par (or larger), of good quality

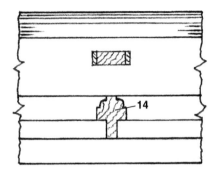

Figure 4.1 (g) *Horizontal section F-F showing* (14) top sash glazing-bar; note the adjacent wedged-end of the mortise-and-tenon joint to the glazing bar of the sash below the meeting rails.

redwood, but preferably of oak or similar quality hardwood;

7.  *Ventilation/weather bead*, 44 × 16mm par (if used, replaces sill staff-bead);

8.  *Back lining* or *backing*, 120 × 6mm plywood (traditionally of sawn softwood);

9.  *Parting bead*, 21 × 9 or10mm par (traditionally ⅜ inch thick);

10. *Wagtail, mid-feather or parting slip*, a 50 × 6mm par length of hardwood or plywood, that hangs securely but loosely from a slot in the pulley-stile head, to keep the boxed sash weights from clashing;

11. *Sash stiles* and *top rails*, 44 × 34mm par (44 × 44 par for thicker sashes);

12. *Bottom sash rails*, 70 × 34mm par (or 94 × 44mm par for thicker sashes);

13. *Meeting rails*, 44 × 28mm par (or 54 × 32mm par for thicker sashes);

14. *Glazing bars*, 22 × 34mm par (or 22 × 44mm par for thicker sashes);

15. *Staff-, Stop-, or guard-bead*, 20 × 16mm par (traditionally ⅝ inch thick);

16. *Glue blocks*, cut diagonally ex 45 × 45mm par;

17. *Sash pulley wheels* (of various metals, sizes and quality);

18. *Sash weights* (ordered in pairs, by the weight of each glazed sash ÷ 2);

19. *Pockets* (formed in pulley stiles to enable access to the boxed weights);

20. *Joggles* are extended and shaped sash-stile horns projecting past the meeting rails that enable the meeting-rail joints to be mortised and tenoned, instead of being dovetailed. The latter was originally done on good-class work.

Of course, it should be mentioned that if a boxframe and sashes are being renewed with a like-for-like replacement, the sizes and style of the original should be followed.

# Multiple boxframe windows

*Figures 4.2(a)(b)(c)(d)(e)*: Double-hung sashes predominate in single pairs, but can be in two or three pairs, with boxed mullions dividing them. Such windows are known technically as *one-light* (one pair of sashes), *two-light*, or *three-light*, according to the number of pairs in each. In three-light windows, when the usually-narrower side lights are fixed

(non-opening) and only the central sashes open, the mullions are solid – not boxed. This was done to lessen the size of the mullion and gain more light through the window. However, the cording and weight-fitting of such centrally-positioned opening sashes is more complex, as the cords have to run across the heads of the fixed side lights to reach the outer boxes, as illustrated below.

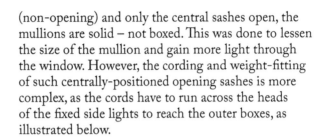

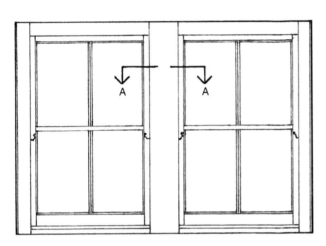

Figure 4.2 (a) *Exterior elevation* of a two-light boxframe window with two pairs of sliding sashes hung on traditional sash cords and weights. Section A-A below shows the detail of the boxed mullion. Note the joggles shown above on the top sashes (which are not only used to avoid dovetailed joints between the sash stiles and meeting rails, but are also reckoned to give better support to the mortise-and-tenoned meeting rails carrying the glass); if you imagine this drawing as being upside down, that would be the interior appearance of any joggles on the bottom sashes.

Figure 4.2 (c) *Exterior elevation* of a three-light boxframe window with boxed outer-jambs (similar to those detailed in Section views D-D and E-E above), two solid mullions, two pairs of fixed (non-opening) sashes and one central pair of sliding sashes hung on traditional sash cords and weights. Section B-B below gives a detailed view of one of the mullions and Section C-C shows the concealed cording-arrangement at the head of the fixed sashes. Note the dotted lines indicating the cording and weighting of the central top-sash.

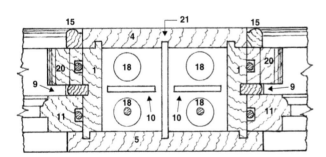

Figure 4.2 (b) *Horizontal section A-A* through the double-boxed mullion in Figure 4.2(a), showing (1) pulley stiles; (4) double inner-lining (166 × 20mm par); (5) double outer-lining (198 × 20mm par); (9) parting beads; (10) wagtails; (11) sash stiles; (15) staff beads; (18) high and low sash weights; (20) joggles (of bottom sashes); (21) box divider (94 × 6mm plywood) – traditionally of sawn softwood.

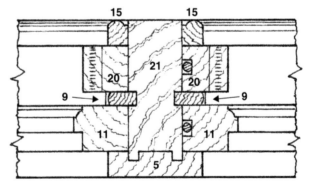

Figure 4.2 (d) *Horizontal section B-B* through a solid mullion in Figure 4.2(c), showing (5) 76 × 20mm par outer lining; (9) parting beads; (11) sash stiles; (15) staff beads; (20) joggles (of bottom sashes); (21) solid mullion (108 × 44mm par).

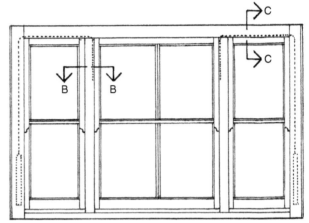

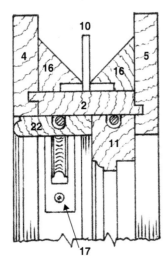

Figure 4.2 (e) *Vertical section C-C through the window-head in Figure 4.2(c), showing* (2) the continuous pulley-stile head; (4) continuous inner lining; (5) continuous outer lining; (10) wagtail; (11) top rail of fixed sash, grooved to accommodate the sash cord from the adjacent top-hung sash; (16) glue blocks; (17) sash pulley wheel; (22) beaded cover-mould, grooved to accommodate the sash cord from the adjacent bottom-hung sash. Note that if standard sash pulley wheels are used, they will require part-housing into the pulley head and a simple modification to the top of the pulley's rear encasement. Also note that renewing sash cords on this type of window involves removal of the fixed (usually skew-nailed) side-light sashes.

## Forming the pockets

*Figure 4.3:* The method of forming pockets in pulley stiles to allow access to the weights varies, but the one detailed below was popular on good-class work. First, it must be mentioned that inserting or removing weights via the access pockets is very difficult if the pockets are too short – and consideration should be given to the potential length of the weights, i.e. the larger the glazed sash, the heavier and therefore longer will be the cast iron or lead weights. Generally speaking, pockets should not be less than 75mm above the top of the sill (to avoid creating short grain at the base of the pulley stiles), nor more than 150mm above the sill (making it more difficult to retrieve sunken weights) – and must not be visible above the meeting rail or joggles of the bottom (inner) sash. About 350 to 400mm is a common length of pocket for small to medium-sized boxframes and – providing these sizes are within the boundaries of the criteria given in the previous sentence, another measure as a final guideline is that the length of pockets should be about a quarter of the height of the pulley stile.

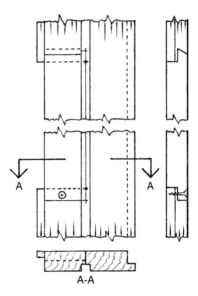

Figure 4.3 Part *front and side elevation and section A-A* of an access pocket in a typical pulley stile. Note that the countersunk screw holding the pocket can also be placed in the saw-cut below, to wedge up the pocket and create a tight fit.

After the pulley stiles have been tongued on their outer edges, grooved for the parting beads and cut to length, the pockets are formed as follows:

1. Mark out the base of a pocket squarely across the inner face-side of the pulley stile, as indicated in Figure 4.3; then square this line across the rebate's edge and step it up on the tongue (into the pocket area) by 12mm and square it across the face of the inner back-side. Determine the height of the pocket and mark this squarely across the face again – but, instead of squaring it on the rebate's edge as before, this time bevel the mark upwards by about 30° and step it down (into the pocket area) by 12mm and square it once again across the face of the inner back-side. Set up a marking gauge from the pulley-stile's edge to the centre of the parting-bead groove and mark this on the face of the back-side, to cut across the squared lines of the pocket. Where these top- and bottom-points of the pocket intersect, drill 6mm diameter counterbored holes carefully with a Jennings' or Sandvik-type auger bit until the point just breaks through into the parting-bead groove. These holes are to relieve the stopped saw-cuts needed and, if not sufficient for inexperienced joiners, larger holes of 12 or 18mm diameter can be made – providing greater care is taken regarding the point of the bit only just breaking through the centre of the parting-bead groove.

2. Still on the back-side of the pulley stile, crosscut squarely and vertically with a gents'-, dovetail-, or tenon-saw into these relieving holes to a depth

equal to approximately half the thickness of the pulley stile (10mm).

3. Turn the stile over to its face-side and – by using the parting-bead groove as a relief area for the stopped saw-cuts – crosscut the square, vertical cut at the bottom and the square, sloping cut at the top to the same half-thickness of stile as before (10mm). Note that these two cuts are the most difficult because the parting-bead groove is usually only 6 or 7mm deep (traditionally ¼ inch) and you may need to counterbore two more 6mm Ø holes in the groove for the final 3 to 4mm saw cuts.

4. Position the pulley stile vertically in a bench vice and, from the face-side, insert a coping-saw blade through the break-through point of one of the relief holes in the parting-bead groove, reconnect the blade to its frame and carefully rip down the centre of the groove until you reach the other break-through point. Although the line of the saw cut should be as straight as possible, providing it is kept within the width of the groove, it will not be disastrous if it wanders slightly in places. Any deviations will be hidden by the parting bead. Note that, if preferred, instead of a coping saw, a jigsaw with a fine-toothed blade could be used.

5. Next, lay the stile face down on the bench, place an offcut of 25mm par timber underneath, just below the lowest cut of the pocket, and give one sharp hammer blow at the base of the pocket. This will break the two small sections of short grain and release the pocket.

6. Finally, plane off the edge-tongue and replace the pocket back in position on the pulley stile, with a central countersunk screw through the bottom lap joint.

## Construction details: drawing the rod

*Figures 4.4(a)(b)(c)*: In Chapter 2, which covers joinery rods, setting and marking out, it was stated that some jobs – like one-off doors – can be made by marking out the components directly, without the need for drawing a rod, laying the components on the rod, marking them off, then marking them out. However, a boxframe window, because it comprises separate parts, is not one of those jobs and – to be able to arrive at the position of the meeting rails (in relation to equal glass-height for each sash), the mortise-positions and shoulder-lines for the sashes, pulley-stile heights, etc – a rod is required that gives a detailed *full-size* vertical- and horizontal-section.

As an example, such (scaled) section views have already been drawn here, related to the boxframe and

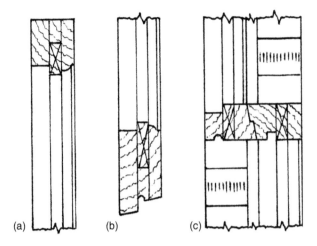

**Figure 4.4 (a)** Size and position of mortise and haunch to top sash rail, as marked on rod; **(b)** similar detail marked on rod for bottom sash rail; and **(c)** mortises marked on rod for meeting rails with joggles. Note that only the vertical position (top and bottom) of the meeting rails would be required if dovetail joints were being used.

sashes in Figure 4.1(a), which was given a 900mm width × 1200mm height × 120mm depth. Therefore, the section views A-A, shown at 4.1(b) and B-B, at 4.1(c), if lined up and set apart to 1200mm, with the meeting rails C-C, at 4.1(d), set midway between them, would be almost what a detailed *vertical section*, (drawn on a rod) would look like. The main differences are that – on a rod – the mortises would be drawn on the ends of the sash rails, as indicated by crosses in Figures 4.4(a) (b)(c) above and the housed pulley-stile height shown. The detailed *horizontal section*, (drawn on a rod) would comprise of section D-D at 4.1(e) and E-E at 4.1(f), lined up and set apart to 900mm, with the glazing bar F-F at 4.1(g) set midway between them. Although there would not be a need to indicate tenon lines on this view, the mortise of the glazing bar might be shown between the edges of the rebate and the ovolo mould.

## Marking off and marking out

If producing a one-off boxframe window, as in Figure 4.1(a), ideally the marking out should be done after the components have been cut slightly oversize to length, planed all round (par), but not yet moulded or rebated. The procedure would be:

1. Position the sill on the rod and – with a sharp, ordinary HB or 2H pencil – carefully mark off the positions of the overall sill-length (which should include the back linings, as these are seated on the sill); then mark the two positions for the pulley-stiles, that are through-housed and side-wedged into the sill (with a 1 in 10 (6°) wedge slope added later on the boxed-in sides of the stiles); then,

finally, mark the two 16mm projections for the edges of the outer linings. Note that these initial *marking off* marks can be short (4 to 5mm) freehand marks or – if you are worried about accuracy – can be made with the aid of a small, 45° plastic set-square squared up from the relevant points on the rod. Either way, this *marking-off* is then *marked-out* squarely to the relevant sides of the sill.

2. Next, lay the pulley-stile head on the rod and mark off its length (which should be to the outer edges of the side-linings, as the plywood backings are usually butted up to the head's underside); then mark off the pulley-stile positions for through-housing – as in the sill, but, this time, without being side-wedged. As before, square these marks across the faces.

3. Position *one* pulley stile on the vertical section of the rod and mark its square-ended length to the pre-marked broken-line *housing depths* (or side-projecting marks that indicate this). Note that the common through-housings in the head should be one-third of the head's thickness (6.66mm, so say 7mm) – and in the sill, where the through-housings are widened to receive thin, 1 in 10 sloping wedges (single, not folding, that logically drive-in from the inner, thick-part of the sill), the stiles should be housed in by an amount that leaves not less than 25mm of sill on the underside, below the housing. If the position of the pockets is already known and set out on the rod, mark the face-side position of these on the stile. Before setting aside, cramp the marked-out pulley stile to the other unmarked stile and transfer the marks squarely across.

4. The lengths of the inner- and outer-linings are usually left with horns on which are removed after assembly of the boxframe – but they can be laid on the rod to mark the positions of the edge-tongues of the pulley-stile head. On good-class work, these tongues are grooved across the back-side faces of the inner- and outer-linings to add to the rigidity and squareness of the boxframe; but on cheaper work, the tongues are removed to avoid the cross-grooving operation.

5. The inner- and outer-lining heads could be marked for length from the rod, but in practise, these are usually marked from, and fitted to, the near-completed frame.

6. A similar procedure is followed regarding the marking off and marking out of the sashes, i.e. shoulder-line marks and mortise-position in the rails for the glazing bars, mortise positions and joggles (if any) related to the height of the sash

stiles – and only single items marked off when there is more than one like-component, to be cramped to the others and used as a rod in itself.

# MACHINING OR HAND-SKILLS PROCEDURES

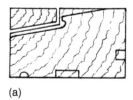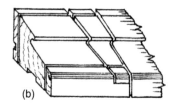

(a)                              (b)

**Figure 4.5** (a) Sill (with a more common, simpler weathered shape) reduced to a manageable, hand-skills' operation; (b) End of sill crosscut to depth for pulley-stile and wedge prior to shaping; and side-ends ripped down to accommodate the inner- and outer-linings either prior to or after shaping – but not removed until the sill is shaped.

*Figures 4.5(a)(b)*: If necessary, the marking out should be completed and the extent of this depends on how much machining and how much hand-skill operations are involved. For example, if the rebates for the glazing were to be done by hand, with a Stanley-type 078 metal rebate plane, marking-gauge lines on the edges and faces would be required. But if the rebates were to be formed with a portable powered router or a vertical spindle-moulder machine, gauge lines would not be required. Likewise, if the tenon-thickness is ripped down by hand (deeped), with a hand saw or a wide-bladed band saw, mortise-gauge lines would be required. But if done on a tenoning machine, gauge lines are not required – except on the end of one rail, for the initial setting up of the tenoning cutters.

Either way, the mortises are usually cut first, as it is more practical to mortise plain, rectangular sections than those that are already moulded and rebated (also, if tenoning by machine, this allows the tenons to be trial-and-error fitted to suit the mortises). If the tenons are not being cut on a tenoning machine (whereby the ovolo moulding would usually be scribed completely through one of the shoulders), the second operation is to rip (deep) them down carefully to the shoulder lines, but without shouldering them (removing the side-cheeks). The third sequential operation is to mould and rebate the components. Then, the side-cheeks of the tenons can be removed with a tenon saw, ready for the scribing of each joint's ovolo mould – as detailed in Chapter 3.

Although sill shapes – especially those with multi-weathered rebates, etc, as shown in Figure 4.1(c) at

section B-B – can be produced by hand-skill techniques, the job is very tedious and demands at least
the use of a fixed circular-saw bench with which
to reduce the sill to a manageable rebated-and-
weathered sawn shape – as illustrated in Figure 4.5(a).
By hand, the sill could then be completed by planing
the weathered slope(s), with such tools as a rebate- or
shoulder-plane (giving relief against any raised steps
for completion with an 05 jack plane), or a bench
rebate plane (which looks similar to a 05 jack plane,
but has a 56mm-wide T-shaped cutter that protrudes
very slightly through small side-apertures to give
a full-width rebate-cut). Of course, if preferred, or
wanting to save time, the approximately-shaped sill
could be finished with a portable electric planer and a
portable powered router.

The end housings in the sill, as seen at 4.5(b), to
accommodate the wedged pulley-stiles, should be
crosscut to depth initially, before any shaping of the
sill takes place – ready for paring down to their gauged
depth after the sill has been shaped. Also, the ends of
the sill can be ripped down initially and – after the sill
is shaped – crosscut and removed to accommodate the
inner- and outer-linings. Similarly, the shallow housings required on the face-side of the pulley-stile head
can be crosscut ready for the pulley-stile housings and
then pared out after the edges have been rebated to
form tongues.

# MEETING RAIL JOINTS AND JOGGLES

*Figures 4.6(a)(b)(c)(d)(e)*: As previously mentioned, the jointing of meeting rails to sash stiles

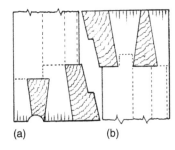

(a)                    (b)

Figure 4.6 (a) Part *side elevation* of sashes showing
the bottom meeting rail of a top sash and the dovetail
arrangement to suit the rail's glass-rebate (shown with a
broken line); and (b) A similar view of the top meeting rail
of a bottom sash, showing the dovetail arrangement to suit
the rail's glass-groove. These joints were usually reinforced
with 6mm Ø dowels through them. Note the 1mm step
between the rail and the face of the sash stiles; this was to
prevent the acute-angled edges of the meeting rails from
scraping against the faces of the stiles and edges of any
glazing bars.

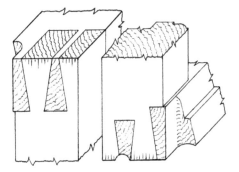

Figure 4.6 (c) The splayed meeting rails (whether rebated
or not) have to be relieved, as shown above, to house the
parting beads. Note that although the relief work is shown
shaped to suit the round-edged beads, it is usually cut
squarely in practice.

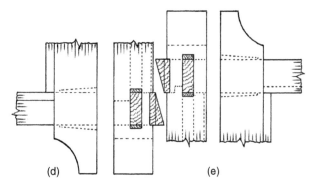

(d)                                      (e)

Figure 4.6 (d) Part *front elevation* and *side elevation* of a sash-corner showing a mortised-and-tenoned,
single-splayed meeting-rail for a top sash with a shaped
joggle; and (e) similar views of a bottom sash – but with
a glass-groove in the meeting rail, instead of the glass-
rebate shown at (d).

on double-hung sliding sashes can be found to vary,
usually according to the period of time in which the
sashes were made. In the Georgian period, through-
dovetails were used on the meeting rails of both
sashes, but this changed to having mortise-and-tenon
joints supported by joggles in the Victorian period.
In the period of time between these two, a mixture of
dovetail joints, mortise-and-tenons and joggles was
used, the former on the bottom, inner sash – the latter
on the top, outer sash. Note that – although shown
here – the rebated splay on the joining edges of the
rails (used to reduce draughts) changed over the years
to an un-rebated splay, shown in illustrations 4.6(d)
and (e).

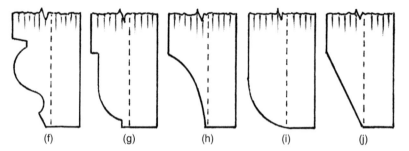

**Figure 4.6** Common joggle shapes drawn to scale at 60mm from the meeting rail; the broken lines indicate the 16mm projection of the outer linings and are used as a guide to position the shape. Note that the 16mm guide also ensures that joggles projecting above the meeting rails of bottom sashes have enough unshaped area to accommodate the sash-cord grooves. The named shapes above are **(f)** ogee, **(g)** ovolo, **(h)** cavetto or scotia **(i)** quarter-round or quadrant **(j)** bevel or splay.

## OUTER-LINING ABUTMENTS

*Figures 4.7(a)(b)(c)(d)*: Where the outer linings are joined to the sill and to the outer-lining head, a few variations can be found and these are highlighted here. Figure 4.7(a) shows the most common recessed abutment of an outer lining to a sill, when the 16mm projection from the pulley stile is notched out to fit over the sill. The alternative at (b) shows the notch scalloped out slightly above the sill, to allow the release of trapped rainwater. At the top end of the boxframe, (c) shows the most common shoulder-line

(un-jointed) abutment of outer-lining head to the outer lining. Such joints are skew-nailed on their top edges with 38mm oval nails and reinforced with glue blocks across the joint on the inside of the box. The alternative at (d) shows a 45° mitred edge – which extends the head by 24mm to allow for the removal of the pulley-stile groove in the outer lining above the mitre, thereby creating a more solid edge to receive the skew-nail fixing. Again, as indicated by broken lines, glue blocks are used to reinforce these butt joints.

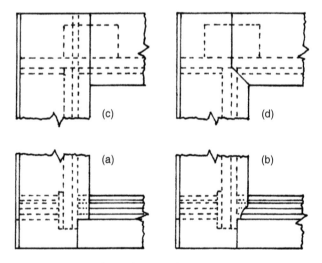

**Figure 4.7** Part *front elevational views* of variations in outer-lining abutments to the sill and outer-lining head:
**(a)** Commonly, the recessed outer-lining is notched over the sill and fitted to the weathered slope; **(b)** The notched projection can be scalloped out to within 3 or 4mm of the pulley-stile cross-housing (as indicated) to inhibit rainwater retention; **(c)** Commonly, at the top of the boxframe, the outer-lining head is butt-jointed and skew-nailed to the outer lining and reinforced (as indicated) with glue blocks; **(d)** Alternatively, the join is partly mitred and glue-blocked.

## VENTILATION/WEATHER BEADS

*Figure 4.8*: As listed in the beginning of this chapter, under component parts No.7, so-called *ventilation/weather beads* can be found fitted to sills as an alternative bottom staff-bead on existing boxframe windows – and therefore might need to be reproduced upon replacement work. As seen in Figure 4.8, these extra deep beads are tongued into the sill and, because of their depth and relative instability, they are sometimes tongued at their ends into the edges of the inner linings also. Their main function for replacing the common staff bead on the sill – apart from keeping out draughts – is to keep out wind-driven rain (which I have seen brimming over a common staff bead fixed to the sill of a boxframe window in a coastal property). These deep beads are also known to have a secondary function, whereby a degree of ventilation at the meeting rails can be achieved without causing a draught; this is done by raising the lower sash and by keeping the deep bottom-rail just below the top of the ventilation/weather bead.

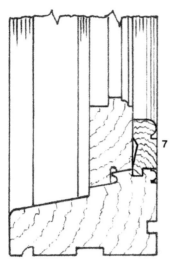

Figure 4.8 A *vertical section* through a typical traditional sill, showing a ventilation/weather bead (7) with a friction-relieving shape against the sash-face. Note that if the ends of this bead are tongued and grooved into the lining on each side, it must be pre-mitred to meet the vertical staff beads and fitted and fixed with the inner linings.

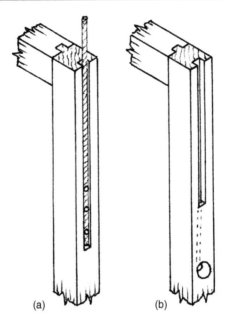

Figure 4.9 (a) Sash cord fixed to a common stopped groove with three large-headed, 18mm long clout nails positioned at about 50mm apart at the lower end of the cord; and (b) A stopped groove with a short tunnel of 9mm diameter at its end, sloping slightly inwards to enter the side of the 22mm counterbored hole below.

## GROOVES FOR CORDS OR CHAINS

*Figures 4.9(a)(b)*: The stopped grooves that are required in the sides of the sashes for housing the sash cord (or chain) are usually 13mm wide × 10mm deep. They can be formed after the sashes have been glued and *cleaned up* (planed and sanded) and they stop at least halfway down the sash. Figure 4.9(a) shows the most common stopped groove, which can be formed with an 050 plough or 045 combination plane (after making a short relief groove by chisel-chopping with a firmer chisel, etc), or more easily with a portable powered router, or other machinery. And 4.9(b) shows a method that can be used for chains or cords – whereby a 22mm Ø hole is drilled to a depth of 22mm and positioned at least 75mm further down from the stopped groove; then (at the end of the stopped groove) a 9mm Ø tunnel is drilled sloping slightly inwards to enter the mid side-area of the larger hole below. The idea being that after threading the cord through the tunnel, a double knot (two knots like conjoined twins) is tied within 25mm of the end, pulled tight, then hammered into the hole; and if used for chains, a steel washer and opened split-pin are attached to the end of the chain and pulled into the large, counterbored hole. Although joiners only usually make and fit the sashes and carpenters hang them after glazing, it must be

mentioned that when fixing cords, only large-headed galvanized clout nails of not-more-than 18mm length (or traditional heavy-duty galvanized staples), should be used. Double, conjoined knots should also be used for attaching sash cord to weights and a steel washer and split pin is one method of attaching chain to weights – if the weights have a counterbored side-hole.

## THE NON-GLUING OF BOXFRAMES

First, it must be mentioned that traditional boxframes were always dry-assembled and nailed – only the sash joints were glued (with non-waterproof *animal*-adhesives, such as Scotch glue), or painted with a lead-pigmented primer and dowelled after wedging, to keep the joints together. Nowadays, totally waterproof, improved PVA-type hybrid resin adhesives, such as EverBuild's D4 Premium Wood Adhesive – conforming to EN 204 and BS 4071 – can be used for interior *and* exterior joinery. My references here, therefore, are for replacement boxframes to be glued.

# ASSEMBLING THE BOXFRAME

After the four pulley wheels have been fitted and screwed into the apertures in the pulley stiles, the sequence of assembly always starts by gluing and wedging the lower ends of the pulley stiles into the sill-housings and then gluing and nailing (with 50mm oval nails) the upper ends into the housings in the pulley-stile head. Next, the assembled inner-frame must be checked diagonally for squareness and sighted from edge-to-edge to check that it is *out of wind* (not twisted). The diagonal check is best done with a so-called *squaring-* or *pinch-stick* with a preformed arrow-pointed end. A length of parting- or staff-bead could be used. With the frame lying outer-side down across a bench, the pointed end of the squaring stick is positioned into an inner corner and the opposite diagonal corner is marked. This procedure is then repeated on the other diagonal of the frame and another mark is made. If the marks coincide, the frame is square. If there are two separate marks, make a central mark between them and force the frame to meet it before proceeding.

Next, because of the relative weakness of the unframed pulley stiles on the pocketed-side, the inner linings should always be fixed first. To achieve this, glue is lightly run along the rebated edges of the tongued pulley stiles (excluding the square-edged pockets!), the pulley-stile head and the recessed housings in the sill – and then the two inner linings and the inner-lining head are quickly positioned, hammered in on a protective *hammering block* and nailed at about 225mm spacing with 38mm oval nails, *punched in* below the surface by about 2mm. The inner-lining head, having been also glued on its shoulder abutments, is then adjusted for flushness and a *blunted* 38 or 50mm oval nail is skew-nailed from the head to the upright lining on each side.

Following this, the frame is turned over on the bench and a similar gluing and fixing procedure is carried out in attaching the outer linings and head.

Finishing operations include 1) placing at least six to eight staggered glue blocks (three or four on each side) in the inner angles of the box's open top, to support and give vertical rigidity to the inner- and outer-linings; 2) inserting the wagtails in the pre-cut slots in the head (note that, as shown in vertical section 4.1(b), at the start of this chapter, a wooden peg or wedge should be inserted through the wagtail to keep it suspended fairly loosely above the boxed sill by about 50mm; whether a peg, wedge or a nail is used, it must be reliably fixed, as collapsed wagtails are irreplaceable in situ and cause the weights to jam up); 3) the plywood back linings are scribed to the approximate shape of the

sill and fixed with 18 or 25mm galvanized panel pins at 150mm centres; 4) the box is cleaned up, usually by *sanding*; 5) the parting beads – which should be a tight fit into the grooves and not require to be nailed – are cut to length and fitted temporarily; 6) the staff beads are mitred and fitted temporarily.

# MAKING AND FITTING THE SASHES

Bearing in mind the theoretical positioning of mortises and tenons – as covered in Chapter 3 – in reality, practical considerations have to be considered and are often given precedence over the theory. For example, the sashes' mortise-and-tenon positions shown in my scaled section views in Figures 4.4(a)(b) and (c) are positioned off centre to suit the 15mm rebate and 9mm ovolo mould required on these stock sizes of sash material. And this also affects the ideal one-third thickness of the mortise and tenon, which has to be reduced from 11.33mm (⅓ of 34mm sash thickness) to 10mm. But then, it is not uncommon to adjust the tenon thickness to suit the mortise.

Whether the sash members have been machine-jointed or hand-jointed, they should be dry-assembled and tested for a good fit. Then, with two sash cramps per sash ready and aided by a glue brush and a damp cloth, each sash is glued up speedily (to beat the initial-setting time of 10 minutes) and cramped immediately. The tenons *and* the inner ⅔ area of the mortises are painted thinly with glue and quickly assembled. Any oozed-out glue should be wiped off with the damp cloth before placing the sash in a pair of cramps and cramping up carefully, close to each joint (with limited pressure to reduce the risk of bowing the stiles). After speedily testing for squareness with a squaring stick, the wedges are glued and driven in. This allows you to release the cramps whilst the sash is lifted and sighted across to test that it is *out of wind*. If *in wind*, place one end of the sash in a vice and strain the other to correct it. Check again and, after removing the twist, wipe off any excess glue and place back in the cramps – if considered necessary. (With relatively lightweight sash material, once the glued tenons are wedged, this usually provides enough mechanical key to hold the joints together until fully set).

Whether you keep the sashes in cramps or not, they should be carefully set aside for a period of time to allow the adhesive to fully set. The manufacturers of the hybrid resin adhesive mentioned above recommend a period of 3 to 4 hours.

After the sashes are cleaned up, the grooves for sash cord or chain are routered out and the sashes are fitted into the boxframes, ready to be hung on site after they have been glazed. Although not the domain of bench joiners, knowledge-wise, the counterbalancing of the glazed sashes involves obtaining separate pairs of sash weights, theoretically with a combined pair-weight equal to the weight of each sash. This is done by weighing each sash with a hooked spring balance and dividing by two. However, in practice, this does not always work and usually each weight for the *top sash* has to be *increased* by 0.25 to 0.5 kg – and *each* weight for the *bottom sash* has to be *decreased* by 0.25 to 0.5 kg. Weighting problems on existing hung-sashes can be caused by re-glazing with lighter- or heavier-weight glass or through a build-up of paint over many years.

# DOUBLE-HUNG SASHES ON SPIRAL BALANCES

*Figures 4.10(a)(b)(c)(d)*: Spiral balances for vertically-sliding sashes were introduced into the industry in the 1940s and a range of these fittings (to suit different sash-weights) are still available from various manu-facturers. Their demand exists for the replacement of existing balances and for use with new sash windows. If using these fittings on sashes in new frames, the frames – as illustrated at (a) and (d) – are usually of solid construction, as opposed to boxframes.

The tubular spiral balances, as shown at (b), can be housed in semi-circular shaped grooves either in the sashes (c) or in the solid jambs of the frame (d). Either way, each of the four tubes containing the spiral balances is fixed to the top of the frame with a drive screw and, after being tensioned with 3 to 5 turns of the fixing plate at the base, is screwed into a 7mm housing on the underside of the sash. Note the top and bottom *limit-stops* fixed in the channels in view (a).

In the case of top sashes with joggles protruding down below the meeting rail (inhibiting the fitting and fixing of the fixing plates at the base of the sash), additional channel fittings are available for an alternative fixing.

# WEATHERING FEATURES

## Sloping sills

*Figure 4.11(a)*: Before completing this chapter with details of other window-types, certain weathering

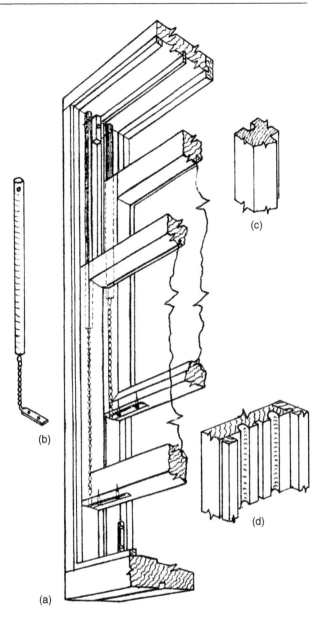

Figure 4.10 **(a)** Double-hung sashes operating on spiral balances in a solid-type frame; **(b)** One of a pair of spiral balances, showing the fixing hole at the top of the tube; **(c)** Part of a sash stile grooved on its side to house a spiral balance tube; and **(d)** Part of a solid window jamb grooved in its sash channels to house spiral balances.

features – already shown here in places – need to be explained. Although common sense perhaps tells us that the slope on the exterior surface of window- and doorframe-sills is to allow rainwater to run off, it will not tell us the most satisfactory angle required. Traditionally, this was reckoned to be about 9° or 10° (1 in 6) or (1 in 5.5), but BS 644: 2003: *Timber Windows – Factory assembled windows of various types*, recommends that such weathering slopes should be

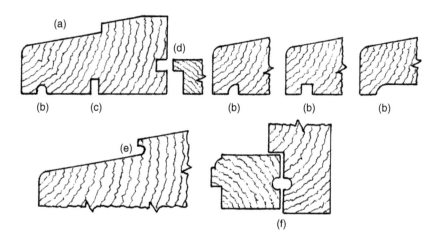

Figure 4.11 (a) A double-weathered slope of 10°; note the alternative, un-throated shallow step between the slopes; (b) These details show four alternative drip grooves; (c) A likely position for a water-bar groove; (d) This groove is for the interior window board; (e) A typical throating groove; (f) Two anti-capillary grooves between a sash and a window jamb.

between 1 in 8 (7°) and 1 in 10 (6°). Interestingly, this is less demanding than the unregulated, traditional criteria.

## Drip grooves

*Figure 4.11(b)*: When rainwater runs off the slope and down the face of a protruding sill, it tends to become overloaded on the edge and starts to seep back (and can be blown back) under the sill. To inhibit this, the underside edge of the sill is grooved with a *drip groove*. When rainwater reaches the outer edge of this groove, it drips off rather than seeping in any further.

## Water bars

*Figure 4.11(c)*: Traditionally, so-called *water bars* (iron or galvanized steel bars with a 25 × 6.5mm section) were grooved lengthwise into the underside centre of sills to inhibit any above-mentioned seepage that may occur. But with the development and escalation in recent decades of modern mastics and sealants (and the simplicity of caulking guns), these grooves and bars are usually omitted nowadays.

## Throating grooves

*Figure 4.11(e)*: On traditional, double-weathered sills (Figures 4.1(c) and 4.11(e)), the shallow, vertical face of the 9mm step in mid-area of the slope (and the rebated bottom rail) was originally grooved out with a *throating plane* – a slim, beech (originally hand-made) grooving plane that was developed to produce these 'throating grooves.' The grooves act like a drip groove in interrupting rainwater seepage.

## Capillary grooves

*Figure 4.11(f)*: These grooves, often opposite each other for maximum effect, can be seen on the side-edges of traditional *and* modern casement windows and doorframes. If being pedantic, they should be called *anti*-capillary grooves. This is because their purpose is to interrupt and break the creeping effect of water seepage between two close surfaces, scientifically known as capillarity – or capillary attraction.

## Casement windows

*Figures 4.12(a)(b)*: Casement windows with solid timber frames and side- or top-hung (hinged) sashes are still being made for so-called *new-builds* and are referred to as *stormproof casement windows* – and in their original form, referred to as *traditional casement windows*, they are still being made as replacements for *old-builds*. The following horizontal- and vertical-sections through the detailed parts of (a), Traditional- and (b), Stormproof-casements will be, hopefully, self-explanatory.

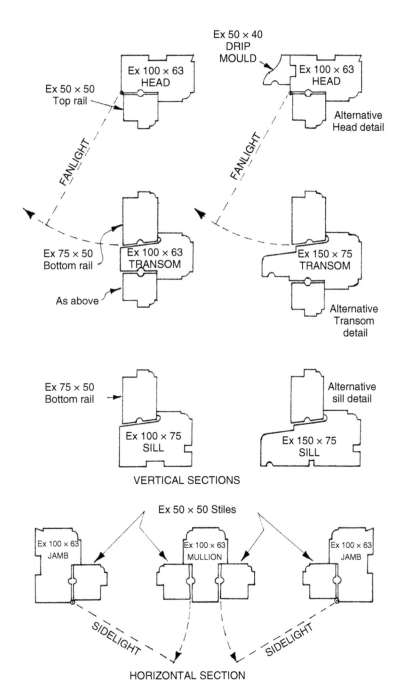

Figure 4.12 (a) Traditional casement window details (proportionately scaled).

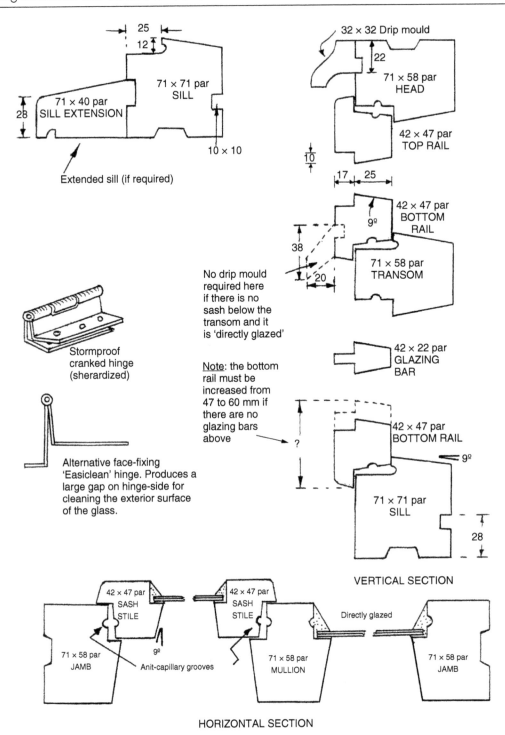

Figure 4.12 (b) Stormproof casement window details (proportionately scaled).

# JOINTING CASEMENT FRAMES AND SASHES

## Sashes

*Figures 4.13(a)(b)*: The frames and sashes for *traditional* casement windows are jointed with mortise-and-tenon joints which, as detailed below, vary in certain detail between *hand-made* and *machine-made*. As shown in Chapter 3, the ovolo-moulded intersections of hand-made joints are partly hand-scribed and – to strengthen the joints – they have haunching spurs and franked haunches. But one of the rail's shoulders on machine-made joints is completely scribed out (by the scribing head on a tenoning machine) to fit the ovolo-moulded shape. The other difference is that haunching spurs and franked haunches do not fit easily into machining operations, so these would be common tenon-haunches mortised into the stiles – as shown at (a).

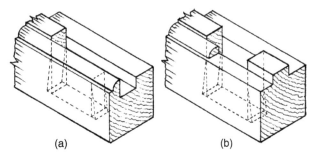

(a)                                (b)

Figure 4.13 (a) Sash stile mortised and haunched to receive a tenon with a machine-scribed shoulder; and (b) Sash stile mortised, ovolo partly removed and haunching spur formed to receive a minimal hand-scribed shoulder. Note the end-grain removal for wedges, which should be done by hand after mortising, to approx one-third of the tenon's thickness, sloping to not-more-than two-thirds of the tenon's depth.

## Frames

*Figures 4.13(c)(d)*: Because the frames are ovolo-moulded and rebated, the techniques of mortising, tenoning and scribing are similar to those described above for sashes. One variation, though, is that the relatively narrow jamb tenons into the sill and the head were left in their full width and not haunched. This was possible because horn-projections were traditionally left on the sills and the head (to be built-in to the brickwork). Nowadays, the horns are removed and these un-haunched joints would require strengthening with draw-bore dowels. Figure 4.13(c) indicates machine-scribing and (d) indicates hand-scribing.

## Hand-scribing technique

*Figures 4.13(e)(f)(g)*: The hand-scribing technique for fitting ovolo- (and other) moulds to each other in right-angled, obtuse- or acute-angled internal corners is simply based on the shoulder-end of the moulding being first mitred (at 45° for right-angles, or other bisected angles for obtuse- or acute-angled intersections). The mitring of a moulded shape produces the elongated profile to be removed by chisel-paring with a scribing gouge and (occasionally) bevel-edged chisels. The illustrations below show the inner shoulder of the tenoned casement-jamb marked and mitred by chisel-paring with the aid of a *mitring template* – and the mitred profile scribed out by vertical paring to about 3 or 4mm below the ovolo-quirk on the jamb's face-edge.

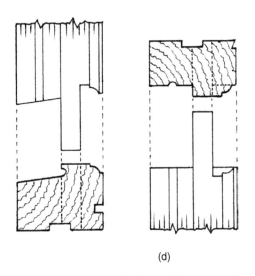

(c)                                (d)

Figure 4.13 (c) The inner shoulder of the tenoned jamb is machine-scribed to fit the ovolo-moulded sill; and (d) the inner shoulder of the tenoned jamb is hand-scribed to fit the ovolo-moulded head. Note the position and size of the mortise-and-tenons: they are governed by the rebate and formed in the thicker part of the components – and do not conform to the third-of-thickness rule. The above mortise-size is scaled at 22mm, an approx quarter of the overall thickness.

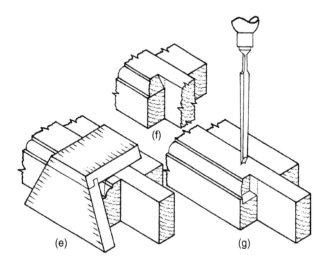

(f)

(e)                                (g)

Figure 4.13 (e) A hand-made hardwood mitre-template (similar metal types are widely available) positioned for mitring the jamb's ovolo mould. This is done by graduated bevel-edged chisel-paring against the template, working back from the top, shouldered corner to the final position shown in the illustration; (f) the completed mitre, ready for scribing; and (g) the scribed area removed by graduated vertical-paring with an in-cannel scribing gouge and a firmer – or bevel-edged – chisel.

## Jointing stormproof casement frames and sashes

*Figures 4.13(h)(i)*: As mentioned above, because built-in horns are not required nowadays (one reason being that their task of securing the frame has been replaced by advanced fixing devices and screwing techniques), it is more sensible for this type of frame and sash to be comb-jointed. Details of this are shown below at (h) and (i).

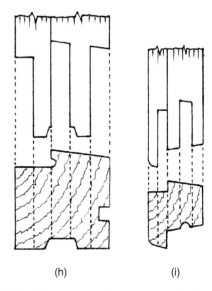

(h)                    (i)

**Figure 4.13 (h)** Part *vertical section* through sill, showing a separated view of the comb-jointed jamb and sill; and (i) a similar view through the bottom sash-rail, showing the separated sash stile. Note that these joints are usually reinforced with one aluminium-alloy star dowel, driven off-centre (to avoid rebates) through each comb joint from the exterior face. These can be obtained in 38 and 50mm lengths.

# HANDMADE COMB JOINTS

*Figure 4.14(a)(b)(c)(d)(e)*: By making short-length wooden templates of the sectional size and shape of the stormproof components, comb joints can easily be marked out and made completely by hand or by a mixture of hand techniques and machining facilities. The sash-stuff's outline can be: 1) marked out with *square-ended* templates (Figures (a) and (c)) onto par square sections to enable the comb's fingers to be set out, gauged and ripped down prior to rebating and splay-moulding, or: 2) the par square sections can be rebated and splay-moulded first, then marked out with *rebated and splay-ended* sash (or frame) templates (b) and (d), to enable the comb's fingers to be set out, gauged, ripped down and shouldered, ready for assembly. The

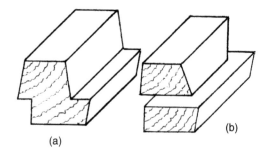

(a)                                        (b)

**Figure 4.14 (a)** A square-ended offcut of sash stuff can be used as a template for marking out the outline shapes and comb-joint positions directly onto the par square sections; and **(b)** Double-ended profile templates can be made from sash stuff and used to mark out the outline-shapes and comb-joint positions directly onto rebated and splay-moulded components.

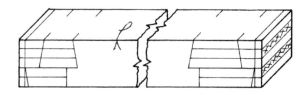

**Figure 4.14 (c)** A stormproof sash-stile marked out on a par component with square-ended templates, showing the division of the 3 + 2 fingers set out and gauged on each side of the rebates – and the waste areas roughly crossed or marked on the end grain to lessen the risk of ripping the tenons down on the wrong side of the gauge lines. Note that the left-hand side of the par component has been marked out with a top-rail/stile template and the opposite end with a bottom-rail template.

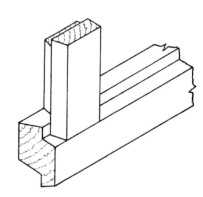

**Figure 4.14 (d)** A sash-stile being marked out for a top rail with a hand-made, double-ended profile template (that can be used to mark top-rails or stiles), positioned (by being slid easily along the glass-rebate) to within 6mm of the end – or to a sash-height mark, then 6mm added for crosscutting.

outer square- and splay-shoulders can be cut with a tenon- or fine back-saw and the inner square- and splay-shoulders can be predrilled near the shoulders and chisel-finished, or cut near the shoulders with

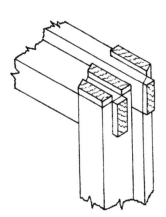

**Figure 4.14 (e)**
*Isometric view* of an assembled comb joint between a top rail and a sash stile, highlighting the minimum waste projections that should be achieved.

a coping saw then chisel-finished. As illustrated at (c) (d) and (e), at least 6mm (square-ended) waste material should be allowed on each of these jointing methods. After the glue is set, this is easily sawn-off and cleaned-up.

# MODERN CASEMENT WINDOW DETAILS

*Figure 4.15*: Modern casement windows with stormproof-type sashes and related weathering details are similar in design to the original *Timber-windows* specification in *BS 644* (Figure 4.12(b)), but they are

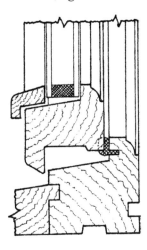

Figure 4.15 Part *vertical section* through the sill and bottom sash-rail of a StormSure casement window manufactured by JELD-WEN to meet the new Building Regulations Part L. The StormSure achieves a range of Window Energy Ratings (WERs) from A to C and whole window U-values of 1.2 to 1.6W/m²K. Note that the 24mm sealed unit uses low-E (emissions) glass; the sill is 68 × 67mm par and the bottom rail is 67 × 55mm par. The un-portrayed jambs and sash stile/top-rail are 67 × 56 and 67 × 55mm par respectively. The weather-seals – as seen in the vertical face of the weathered rebate above – are Schlegel's AquaMac 109.

given much wider rebates to accommodate double-glazed sealed units, are made air- and water-tight with AquaMac-type weather-seals and are either hinged traditionally, or are more sophisticated with concealed, side-fixed *geometry friction hinges*. Some designs allow the sash to be fully reversible for interior maintenance and window-cleaning. And, because of the more substantial timber sections required to accommodate the double-glazed sealed units, even open-and-shut trickle vents can be fitted in the sash's top rail or the casement frame's head.

The timber-beaded, sealed-unit rebates can be internal or external – I prefer the former, but the latter seem to predominate. External beads are usually pinned and internal beads can be pinned or screwed – if the windows are made of hardwood and internally beaded, they should be cup-screwed. A local glazing company in my hometown recommend bedding sealed units into timber sashes with butyl glazing-compound in preference to double-sided glazing-tape methods – but, again, the latter seem to predominate.

# UPGRADING TRADITIONAL-TYPE CASEMENT WINDOWS

*Figures 4.16(a)(b)*: As per paragraph 4.18 of Part L1B's Guidance notes, the *upgrading* of existing windows is not notifiable to any accredited- or

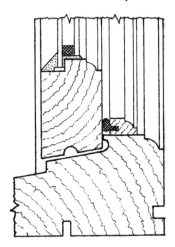

Figure 4.16 (a) Part *vertical section* through the sill and bottom sash rail of an upgraded, traditional casement window, showing a 16mm double-glazed sealed unit with stepped edges resting on 2mm-thick plastic 'setting blocks'. The unit is held with glazing sprigs and bedded and face-pointed in butyl glazing compound. Additional upgrading is achieved by setting a Schlegel's AquaMac 109 or 63 compression weather-seal into the narrow-grooved 20 × 10mm ovolo-moulded beading, mitred and fixed around the inner faces of the frame.

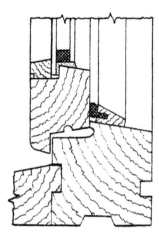

**Figure 4.16 (b)** Part *vertical section* through the sill and bottom rail of an upgraded, traditional stormproof-type casement window, showing a 16mm stepped, sealed-unit and splay-beaded AquaMac 109 or 63 compression weather-seals mitred around the inner splayed faces of the frame; all similar to Figure 4.16(a), except the front timber glazing-beads – the inner-edges of which are either faced with double-sided butyl glazing tape, or butyl glazing compound.

local-authority and does not have to comply with Part L's standards regarding *the WER of the whole window and frame*. Even so, the energy rating of existing windows can be improved by removing the single-thickness glass and re-glazing the sashes with double-glazed sealed units with made-to-measure stepped edges. Also, it is usually possible to fit weather-seals around a sash, although this involves forming very narrow grooves in the edges of purpose-made timber beading; this can be done with special weather-seal router cutters (on full-width boards, before cutting to bead-thickness). The beading is mitred and – as illustrated – surface-fixed around the frame's interface with the sash.

# UPGRADING TRADITIONAL BOXFRAME WINDOWS AND DOUBLE-HUNG SASHES

*Figures 4.17(a)(b)(c)*: The upgrading of weighted double-hung sashes presents more problems than upgrading traditional casement windows. This is because the sashes are counterbalanced with the corded sash-weights in the boxframes – and when the single-thickness glass in the sashes is replaced with double-glazed, stepped sealed-units, the sashes are heavier and have to be re-counterbalanced with

new weights (as described above, under *Making and Fitting the Sashes*). Although new *lead* weights are sometimes used to increase the weight, if they need to be longer than the originals, the length of cord is shorter and lessens the amount of travel and opening area.

Alternatively, the sashes could be modified to receive patent spiral-balances; but this would involve them being removed, de-glazed, grooved on their side-edges, re-glazed with stepped sealed-units, then weighed to determine the weight-type of balances required. To overcome the logistics of such a task (brought about by retaining the existing boxframes), the Ventrolla Sash-Window Renovation Company takes site measurements and makes and installs new sashes with pre-glazed sealed units. The replacement sashes are grooved and fitted with *pile-carriers* and *Weatherfin pile* on the abutting edges of the top- and bottom-rails and between the meeting rails, on the splayed edge of the bottom sash. On site, their installers renew the parting- and staff-beads – the latter grooved and fitted on the inner edges with poly-propylene pile-carriers containing Weatherfin pile – and the parting beads being plastic replicas with a strip of Weatherfin pile fitted on one side, which must face towards the exterior. A 'U' shaped polypropyl-ene section is screwed into the original parting-bead groove and, as illustrated at (b), the detachable pile-carrier bead snaps into this and allows future removal for sash and cord maintenance.

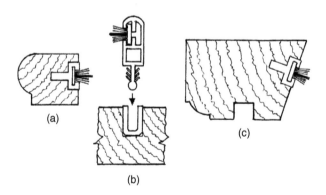

**Figure 4.17 (a)** *Critical parts of the Ventrolla Perimeter Sealing System* (VPSS), showing a carrier channel and Weatherfin pile fitted into the inner edges of their 'small' staff bead; and **(b)** a uPVC replica parting-bead fitted with a Weatherfin pile, poised to snap into the modified channel in the pulley stile below; and **(c)** the meeting rail of a bottom sash also fitted with a carrier channel and Weatherfin pile. Note that the polypropylene fin protruding in the centre of the silicone-treated pile, flexes over to an 'L' shape when compressed between the normal sash movements.

# 5
# Making doors and doorframes

## INTRODUCTION

The making of doors – as with windows – is nowadays mostly in the domain of large manufacturers using CNC (computer numerical control) production processes and CAD/CAM (computer-aided design and manufacture). Also, *engineered composite-* and *architectural fibreglass-doors* have a sizeable slice of the market. However, small- and medium-sized joinery shops do still get asked to make bespoke timber doors, be it mainly for exterior types, including shaped-doors (Tudor-headed, etc) and one-off, non-standard size internal doors. Regarding purpose-made exterior doors (and their frames), it must be mentioned that these should comply with the amended Building Regulations' AD (Approved Document) Part L1A and L1B: *Conservation of fuel and power in new and existing dwellings*, regarding achieving the current U-value of $1.8W/m^2K$ to meet the DER (*dwelling's energy rating*).

## COMPLIANCE WITH AD, PART L1B

Although any upgrading to the insulation of dwellings is mutually beneficial (to the Government via the Building Regulations addressing the environmental issues and to the occupiers of a property, reducing their energy bills and their carbon footprint), such regulations were not previously enforceable retrospectively on existing dwellings. However, the amended Part L refers to exterior doors and windows as 'thermal elements' that – *if replaced* – are regarded as *controlled fittings* (controlled by having to comply with the upgraded energy rating). But the obligation to do so, only applies to *a whole unit*, i.e. including the doorframe. The replacement of a door (or a window) into an existing, retained frame, therefore, does not have to meet the Part L standards – although, where possible, it would be sensible to do so.

## UPGRADING

There are at least five ways to upgrade new exterior timber-doors and doorframes to achieve lower U-values: 1) by pre-grooving the rebated inner-edges of the frame to receive *compression* weather-strip seals (illustrated further on in the chapter); 2) by fitting and fixing a good-quality patent weather-seal/draught excluder to the frame's sill; 3) by achieving a finished door-thickness of at least 45mm; 4) by using rigid insulating-material such as sheet cork or Celotex insulation board snugly sandwiched between double panels; and 5) by glazing any fixed-light areas with 24mm sealed units using *low-E* (low emissive) glass and *Warm-Edge* spacer bars.

## BASIC DOOR KNOWLEDGE

Basically, doors are usually referred to by standard sizes and the number of panels or glazed areas that they contain. And those without panels (which are covered with plywood or hardboard) are usually referred to as *flush* doors. Combined references might also be applied, such as *half-glazed flush door*, or *four-panelled door with raised-and-fielded panels*. We must also know the difference between *internal* and *external* doors. The latter – apart from usually being slightly wider than internal doors – have to be strong enough to provide security and withstand extreme weather conditions on one side, yet not lose their balance from the different conditions on the other – so they are usually thicker, therefore heavier, and of a more substantial construction than internal doors.

More importantly from a structural viewpoint, doors – some quite heavy, as mentioned above – hang free from their hinged side and the framing arrangements have to withstand this suspended load. In the case of *panelled doors*, the framing strength is achieved by the jointing of the horizontal rails to the vertical stiles, which creates numerous structural right-angles (like un-braced gallows' brackets) – the most supportive ones being the inverted *brackets* created by the wider rails, such as the bottom rail; this being the main reason why it is wider than the top- or intermediate-rails. Wide middle rails also add extra supportive strength, but their main original function was to accommodate a mortise lock – hence they are also called *lock rails*. In the case of *flush doors*, the interior perimeter-framing and cross-rails can be of much smaller sections, because the framing strength is gained when the outer layers of plywood (or hardboard) are glued on, thereby creating a self-supporting unit. Finally, in the case of *framed, ledged, braced and matchboarded (FL&B) doors* and *ledged, braced and matchboarded (L&B) doors*, the framing strength is achieved via the diagonal bracing between the horizontal ledges – where the gallows' bracket principle is openly displayed.

## BASIC DOOR TYPES

Basically, there are five types of door and bespoke, purpose-made joinery nowadays is only usually involved with the first three types. These are 1) panelled doors; 2) glazed doors; 3) FL&B- and L&B-doors; 4) flush doors; and 5) fire-resisting (FR) doors.

## PANELLED DOORS

*Figures 5.1(a)(b)(c)(d)*: The number of panels depends on design, but they are usually between two and six. One-panelled doors are feasible (and do exist), but if the panels are of solid, edge-jointed timber, the rails and stiles need to be of quality redwood or hardwood of a substantial size. Also, the stiles should be wider and the bottom rails should be deeper (by at least 13mm) than the norm of ex 100mm for stiles and ex 200mm for rails. And because of the lack of extra support from middle- and intermediate-rails, the dry-jointed panels should only be given a few millimetres fitting allowance *in width* (for expansion and contraction), *not in length* (height). The thermal movement of converted timber lengthwise is scientifically known to be negligible, so such a snug fit of the panel between the top- and bottom-rails will add extra supportive strength to the framing arrangement. As illustrated at 5.1(a) and (b), the most common number of panels seems to be four and six.

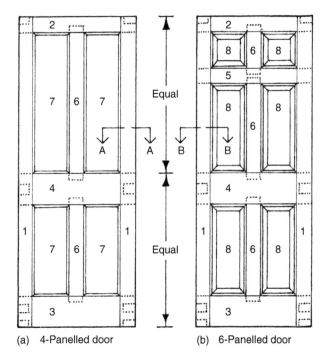

(a)   4-Panelled door          (b)   6-Panelled door

Figures 5.1 **(a)** and **(b)** *scaled elevations* of panelled doors; the broken lines show the hidden detail of the mortise-and-tenon joints and the *horizontal section lines* **A-A** and **B-B** refer to details shown separately. The numbered parts are: 1) stiles; 2) top rails; 3) bottom rails; 4) middle- or lock-rails; 5) intermediate rail; 6) muntins; 7) plywood or solid timber panels; and 8) raised, sunk-and-fielded panels. Note that – although not shown – the above through-tenons should be wedged – and the muntins' stub-tenons (in practice) are not, (although some text books show them fox-wedged).

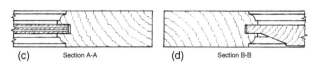

(c)   Section A-A          (d)   Section B-B

Figure 5.1 **(c)** 9mm × 9mm Ovolo-moulded door stile (of interior-type door) grooved (12mm deep) to receive 9mm-thick plywood panels; and **(d)** similar grooved and moulded door stile with a one-sided raised, sunk-and-fielded solid panel. Note that when the tongued edges of the raised panels are being formed – whether by hand or machine – a 100mm to 150mm long, purpose-made grooved block (traditionally called a 'mullet' and resembling an offcut of grooved stile) can be slipped along the edges to control the snugness of the fit.

### Scribed-joint details

*Figures 5.1(e) to (n)*: Scribing and fitting edge-moulded components has already been illustrated and described in Chapter 3 and Chapter 4, but Fig.1(e)(f)(g) and (h) below highlight points of difference between *hand-made joints and scribes* and *machine-made joints and scribes* (indicated with broken lines in Figures 5.1(a) and (b) above). Figure 5.1(e) below shows the part-removal of the ovolo mould beyond the 6mm gouged-out scribe by hand-scribing of the top rail and 5.1(f) shows the machined scribe running through the width of the top rail, leaving the ovolo-moulded edge mostly intact. Note how the haunching-depth differs between hand-made and machine-made joints, with regard to the latter's haunch being accommodated in the panel's groove.

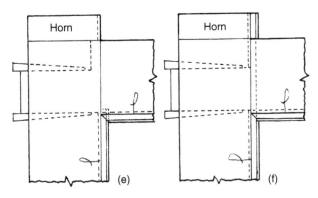

Figures 5.1 (e) and (f) Part *elevational views* of hand-scribing (e) – and machine-scribing techniques (f) of the mortise-and-tenon joints between the top rail and stile of the door at 5.1(a) or (b) – as described above.

As illustrated at 5.1(g), because the 9mm-wide panel groove is narrower than the 12mm-wide mortises in this 35mm-thick interior door, two 1.5 × 12mm slivers of moulded edge (*) have to be removed (pared off with a chisel or craft knife) on each narrow edge of the tenons – as do two 1.5 × 4mm (quirk-depth) slivers of beaded edge (*) on machine-scribes and the whole mould on hand-scribed joints.

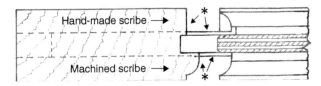

Figure 5.1 (g) Part *horizontal section* through the ply panelled door, showing a partly open mortise-and-tenon joint between the stile and a panelled rail; in reality, this joint could only fit together if the slivers of tenon-edge and mould (*) were removed, as at (h). Note: hand-made and machine-made joints are not mixed as in these illustrations.

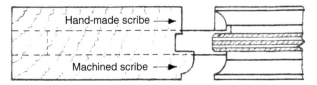

Figure 5.1 (h) The removal of the 1.5 × 12mm deep slivers on the narrow edges of the tenon can be visualized by comparing 5.1(h) with 5.1(g); the removal of the 1.5 × 2.5 mm slivers of face-edge mould and the hand-scribed shoulder-mould (in the joint area) are shown as already removed at 5.1(g) and 5.1(h) to avoid confusion. Note: for greater clarity of the one-dimensional joinery detail given above, compare the information with the three-dimensional isometric views below at 5.1(i) to 5.1(n).

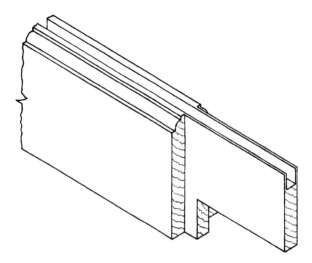

Figure 5.1 (i) The 1.5 × 12mm groove-slivers to be removed from the top rail's tenon.

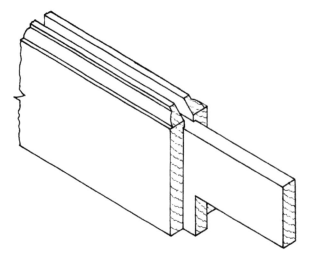

Figure 5.1 (j) Slivers removed and ovolo moulds mitred to give profile of scribes.

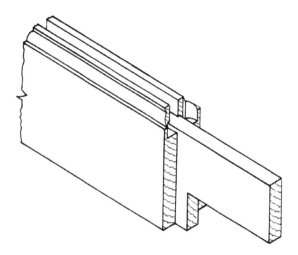

Figure 5.1 (k¹) Ovolo scribes gouged out 5 or 6mm beyond the shoulder-line quirk.

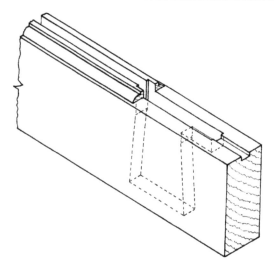

Figure 5.1 (m¹) The ovolo mould has been chisel-chopped and pared off to create the shoulders for the hand-scribed jointing of the top rail. Note the remaining horn.

Figure 5.1 (k²) Example of ovolo-moulded and grooved rail tenoned and scribed by hand.

Figure 5.1 (m²) Ovolo-moulds initially removed by chisel-chopping prior to cross-paring down to the quirk-line on each side.

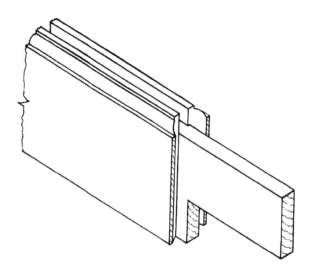

Figure 5.1 (l) Machined ovolo scribes continuous across each shoulder. The 12mm haunch projection equals the tenon-thickness and the depth of the panel-groove.

Figure 5.1 (m³) Ovolo-moulds pared down to the quirk lines (now shoulder-lines) on each side.

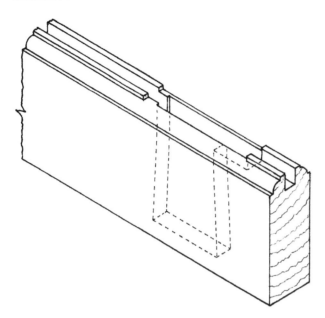

**Figure 5.1 (n)** The 1.5 × 2.5mm slivers of mould-edge have been pared off in the area of the mortised stile to receive the machine-scribed top rail.

**Figure 5.1 (p)** The assembled (upside-down joint) resembling a part-stile and part-top rail of an ovolo-moulded door, grooved to receive 9mm panels; note that – for this example – the haunch was omitted.

## Preparation of door material

When crosscutting the various rails to size, they should have about 20mm added to their length. This additional amount is to give the tenons a 10mm projection at each end. There are four good reasons for this, as follows:

1. When deeping (ripping or deep-sawing along the grain lengthwise, parallel to the widest face of the timber), with a hand-saw to produce the cheeks of the tenons, sometimes the initial saw cut jumps on the end grain and can be a bit erratic until the cut is established after a few strokes. By having a waste-projection, any unwanted saw-cut marks will not matter;
2. When eventually gluing up, the wedges can be more easily held against the edges of the projecting tenons, to facilitate being driven in squarely;
3. A reasonable-sized projection makes it easier to saw off the waste material after gluing and wedging, leaving a fraction to be *cleaned* (planed) off;
4. The addition on the tenons gives more length of waste in the haunch area, from which to cut two

**Figure 5.1 (o)** Assembled, open joint, giving a view of the shoulder related to the scribing.

or more wedges – as illustrated in Chapter 3. Note that wedges cut like this, from the tenon's waste area, are the exact required thickness.

As illustrated in Figures 5.1(e) and (f) above, about 60 to 80mm more than the door's height should be added to the initial length of the stiles to produce projecting *horns* of 30 to 40mm at each end. The main reason for the horns is to add more length to the *short grain* above the outer wedges and thereby increase its resistance to shear when the wedges are being driven in. Traditionally, a second reason for having horns was to protect the vulnerable outer corners of doors during transit. Also indicated in Figures 5.1(e) and (f) are typical looped *face marks*, usually made in a few places with a pencil and joined up on the timber's edge with an open arrow mark. These symbols are made to remind the joiner/machinist of the selected best face and edge. Therefore, they are named *face-side* and *face-edge marks*. They are applied after machining and after the various pieces of timber have been scrutinized. This involves studying the characteristics of the grain, the position of any knots or blemishes and the position of sapwood in relation to heartwood; the latter being chosen, if possible, as the outer face on exterior doors.

## Selection and preparation of timber

If the timber is in a sawn state, as is usual in joinery workshops – it must first be selected and cut to initial lengths with either a crosscut- or hardpoint-saw, a portable, powered crosscut/mitre saw, a portable circular saw, or a travelling-head crosscutting machine. Then,

to produce the sawn, sectional sizes (which usually have at least 6mm added to the finished sizes for planing), it will need to be deeped and/or *flatted* (flat-sawn along the grain lengthwise, parallel to the face-edges of the timber) with either a narrow bandsaw machine (using the widest blade available to avoid *snaking*), or a realistically-robust circular saw-bench machine.

## Machine planing

*Figures 5.2(a)(b)*: The next job is planing – and although hand planes are still used occasionally by joiners, the task of *planing* timber *all round (par)* by hand is of a bygone age – and therefore a surface-planer/thicknessing machine would be used. Although training should be acquired and health-and-safety regulations adhered to, the sequence of using this machine is as follows: Minimise the exposed, cylindrical cutter-block that lies just below the open-surface of the machine's top bed, by adjusting the fence to suit the timber's width and adjust the bridge guards (one on each side of the fence) to allow a slight clearance for the sawn timber's thickness to pass under. Then plane the wide face-sides of each piece until all saw-marks are removed. As illustrated in Figure 5.2(a), this includes feeding the timber smoothly, with *light pressure* and slow speed; the hands should be positioned as shown, on each side of the bridge guard (the left hand having been lifted over the guard to hold down the emerging timber), then – near the end of a planing operation – the right hand is moved over the bridge guard.

**Figure 5.2 (a)** Hand-positions and stance for planing the 'face side'.

Figure 5.2 (b) Hand-positions and stance for planing the 'face edge'.

Next, the machine must be isolated (the power switched off) and, with a metal try-square, the fence should be checked and adjusted for precise squareness to the machine bed. When squared, place a length of the surfaced timber against the fence, drop the bridge guard to its lowest position and adjust it laterally to just clear the sawn face of the material. Then remove the test piece, switch on the power and start machining the face-edges to the stiles and rails, as illustrated in Figure 5.2(b).

Whilst passing the pieces over the revolving cutter block, unrelenting *light hand-pressure* must be applied to keep the face-sides against the fence, ensuring squareness of the edges – but, more importantly, too much feed-pressure can cause the timber to snatch and fly back, leaving the cutters exposed to the operator's over-exerted downwards thrust. This safe practice is achieved by keeping the outstretched fingers of both hands pressed against the sawn-face and the thumbs in a *thumb-printing position* on the timber's top edge. To accomplish this, you will have to move (after the initial start) to the side of the machine, facing the end of the cutter block. Note that – even though the cutter block is covered by the bridge guard and/or the timber being machined – you must develop the safe practise of lifting the relevant hand off the timber when passing over the cutter block. And never hold your clenched fingers *behind* material being surfaced or edged when passing over the cutter block.

After surfacing and edging and isolating the machine again, the bridge guard must be readjusted to completely cover the cutter block and the under-table

is wound down to the machine's depth-gauge measurement to remove the first few millimetres of the surplus width. This allows the material to be checked for size before the final setting of the rollered undertable to the required finished width. Finally, a similar sequence of operations is repeated to obtain the finished thickness of the material. Note that thicknessed widths precede thicknessed thicknesses to help maintain stability of the relatively slim timber-on-edge passing through the pressurized underbelly of the machine – which must only take one piece of timber at a time (unless the machine has a sectional or recessional feed-roller) and the timber's length must not be less than the measurement between the infeed and outfeed rollers. Usually, 300mm is a safe minimum – but individual manufacturer's instruction literature should be carefully and fully read.

## Hand planing

*Figure 5.3*: Planing sawn timber all round by hand, using a metal jack- or try-plane, requires a good degree of developed skill to achieve a true, flat surface, square edges and parallel width and thickness – to the sizes required. However, planing-all-round, if practised is an excellent introduction to the basic skills of general hand-planing. The procedure is as follows:

● Select one of the wide surfaces of the sawn timber to be the face-side and – simply expressed – plane off the rough surface. However, more precisely, the planing has to be controlled: 1) to achieve *surface-straightness* (with no *rounds* or *hollows*) in

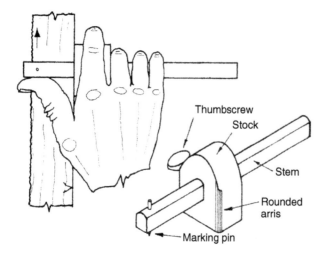

Figure 5.3 Hand-position on gauge for marking the finished thickness.

*length*, by sighting along the edges and/or check-ing the surface with a straightedge; 2) to achieve *an untwisted surface*, by sighting across a pair of pre-positioned *winding sticks* (small, parallel straightedges); 3) to achieve *flatness* (with no *rounds* or *hollows*) *across the material's width*, tested by tilting the base of the metal plane across the surface occasionally, to highlight any inaccuracies below its edge; and 4) to achieve *a surface clear of chatter marks and broken-edged cutter marks*, by applying enough pressure to the moving plane and lifting its heel at the end of a forward thrust to *break* the shaving. Note that when planing at the timber's end nearest to you, keep more pressure on the *front* of the plane (and be careful not to draw back too much and come off the end!) – and when finishing a plane-thrust at the timber's end farthest from you, keep more pressure on the *back* of the plane until the plane's cutter has cleared the end of the timber. Next, apply a few pencilled *face-side marks* towards your selected best edge.

- Now place the timber in the vice and plane the face-edge to be *straight* (by sight and/or straight-edge *and square* to the face side (by checking with a try-square occasionally, pencilling the left- or right-hand side high-edge points and removing them by shifting the laterally-central position of the plane (the default position) to one side or the other). When true and square, apply pencilled V-shaped arrow-heads as *face-edge marks*.
- Next, set up a marking gauge to the finished width and gauge each side. Cramp the material in the vice and plane carefully (meaning vigilantly) down to the gauge lines.
- Finally, reset the marking gauge to the finished thickness and gauge each side (as indicated in Figure 5.3 above). Then lay the material on the

bench, against the bench-stop, and plane carefully (vigilantly) down to the gauge lines.

## Using the marking gauge

For a right-handed person to use the gauge, it should be held as shown, with the thumb on the stem (behind the pin), the forefinger resting on the semi-circular surface of the stock and the remaining fingers at the back of the stock, giving side pressure against the timber being marked. Always mark lightly at first to overcome any grain deviations. Choose a manage-able amount of material to gauge (say 300 to 500mm) close to the front end; gauge it lightly, then more heavily, then move backwards from the end of the gauged line and gauge another manageable amount; and so on until the rear end is reached. This technique might sound disjointed, but in experienced hands, it flows almost seamlessly. The gauge is easier to hold if the face-edge arris – that rubs the inside of the out-stretched thumb – is rounded off as shown.

## Marking out the door

Whether machine-planed or hand-planed, the next operation is to *mark out* the stiles and rails, etc, with the exact length, width and joint details – this was covered in Chapter 2.

## Shaped-headed doors

*Figures 5.4(a)(b)(c)(d)(e)*: The common geometrical shapes for shaped-headed doors and their frames (covered in Chapter 8: Geometry for Curved Joinery) are *1) segmental; 2) semi-circular; 3) semi-elliptical; 4) Tudor;* and *5) Gothic*. Whether these doors are for interior or exterior use has to be taken into account, but the main issue for the joiner – dealt with here – is how best to form the shaped heads and join them to the stiles. Four of the five door-head shapes named above are illustrated in Figures 5.4(a) to (e), related to panelled doors. A semi-elliptical headed door is not shown because of its similarity to the segmental-headed door at (a).

## Jointing details

### Segmental-headed door

*Figure 5.4(a)*: As illustrated, the bottom- and middle-rail joints are common mortises-and-wedged-tenons, but the segmental top-rail has to be jointed differently. I have indicated open-topped stub tenons and such joints should be strengthened with draw-bore dowels after the door has been glued and cramped. As an

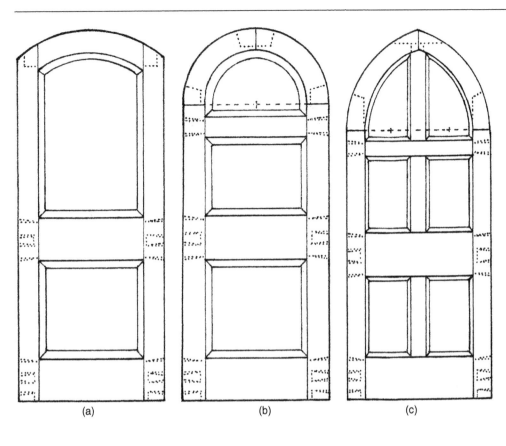

Figure 5.4
(a) Segmental-headed, 2-panelled door with raised-and-fielded panels; (b) semi-circular headed, 3-panelled door with raised-and-fielded panels; (c) depressed-Gothic headed, 6-panelled door with 16mm panels and 'stuck' (see definition to follow) mouldings. Note the 50mm needed between springing line and intermediate rails.

(a)        (b)        (c)

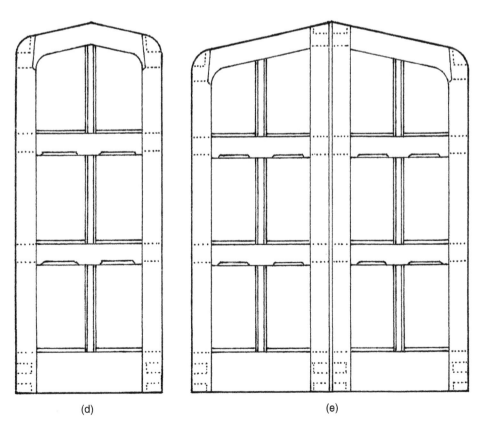

Figure 5.4 (d) Straight-top Tudor-headed, six panelled door with 18mm solid panels, bevelled and stop-chamfered intermediate rails and bevelled muntins and bottom rail; (e) straight-top Tudor-headed double doors of a similar design and construction to (d), but with bead-moulded and rebated meeting stiles. Although not indicated above, the muntins would be stub-tenoned and the through-tenons would be wedged.

(d)        (e)

exception to the norm, G-cramps should be applied to the cheeks of the open-topped tenons, in addition to normal cramping with sash cramps. However, the straight shoulder-abutments, as shown, would only be possible if the shaped rail had been formed by gluelam bending. As explained in Chapter 3, this is when thin laminae (strips of timber) are glued and cramped together in purpose-built formers to produce curved shapes. Otherwise, if the curved top rail was formed by being cut to shape from one piece of solid timber (a traditional method), the straight shoulder-abutments would create short grain at the ends of the concaved shape, so – as shown in Figures 5.4(d) and (e) – the shoulder abutments should be mitred and splayed.

## Semi-circular headed door

*Figure 5.4(b)*: As with all shaped-headed doors, there are alternative ways to shape the heads and alternative methods of jointing and joining the heads to the stiles. Again, the shaping shown here refers to two quadrant-shaped segments cut from pieces of solid (or built-up – laminated) timber, jointed at the crown and to the stiles on the springing line. The springing line is about 50mm above any intermediate rail. This is done mainly to promote an easier transition from curve to straight and partly to create two horns to strengthen the mortise-and-tenon jointing of the intermediate rail. Radial-ended tenons are shown on the stiles and a radial-ended loose-tenon insert is indicated at the crown. The technique of dowelling described for the first door should also be used on these three joints. Traditionally, *hammer-headed tenons* or separate hardwood *hammer-headed inserts* would have been used on the three joints in question, but they involve an enormous amount of work and hand-skills.

## Depressed-gothic headed door

*Figures 5.4(c) and (f)*: Again, there are alternative ways of shaping and jointing the two arched stiles that form the head of this type of door. My own idea (to add strength to the relatively-weak top-heavy arched third of the door) is to make the top muntin continuous with the middle muntin, passing as one component through the intermediate rail and forming a joined-up, self-supporting cross. Thereby, instead of relying wholly on side-support from the springing-line joints, the arched stiles would also be supported by the muntin's stub tenon into the underside of the crown joint. To make the cross, yet keep the appearance of the two muntins being separated by the horizontal intermediate rail, it could be built-up with three gluelam thicknesses: With a door thickness of 45mm, the tenon-thickness would be 16mm (equal to the panel thickness) and this would be the thickness of the

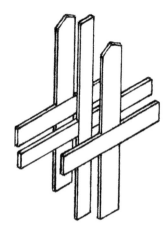

Figure 5.4 (f) An exploded isometric view of the built-up gluelam cross.

sandwiched, square-edged gluelam components (one vertical and two side arms). The outer, edge-moulded laminae on each side of the central, tenoned cross would be 14.5mm thick. To make more sense of this explanation, see Figure 5.4(f) above. The other joints are similar to those indicated for the semi-circular headed door in Figure 5.4(b).

## Straight-top tudor-headed doors

*Figures 5.4(d) and (e)*: The top rails of these attractive Tudor-period doors could be partly (and economically) built-up on the underside of their joint-ends for shaping – and, as illustrated, the shoulders should be part-mitred and splayed to avoid 'feathered' edges and short grain. On exterior doors, the bevelled top edges of the rails, shown here, would of course act as weathered edges.

## Methods of constructing heads of shaped-doors and/or frames

*Figures 5.5(a)(b)(c)*: Apart from shapes cut uneconomically from diagonally-orientated solid pieces of timber, illustrated at (a), which can create problems with short grain, the alternative methods of forming curved shapes are shown at (b) and (c) below:

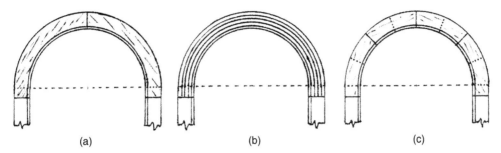

Figure 5.5 (a) This semi-circular construction is formed with two solid-timber quarter-turns, with their grain tangential to the curves (as indicated) and can be jointed at the crown with a loose tenon (or a double tenon) insert and at 50mm below the springing line with single or double jamb- or stile-tenons; (b) shows a semi-circular door- or frame-head made-up of gluelam laminae with a thickness equal to the semi-circle's inner radius divided by 150; the shape is glued and set up in a purpose-made male and female former; and (c) is of a similar construction to (a), but the short grain and the timber wastage is less severe; structurally, the side-thickness of the shape is made up of two layers of four pieces on one side and five on the other, with the radial joints staggered on each side. On one side the joints are positioned at 45°, 90° and 135° and on the other side (shown dotted) they are at 22.5°, 67.5°, 112.5° and 157.5°.

## Exterior frame details

*Figures 5.6(a)(b):* Whether of softwood or hardwood, an exterior frame should be of a substantial width and thickness, usually not-less-than 95 × 57mm par (finish), from which rebates of at least 12 × 46mm should be removed to house the door.

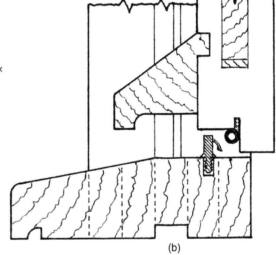

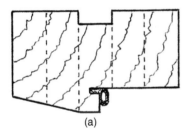

Figure 5.6 (a) A *cross section* through a 95 × 57mm par doorjamb and/or frame-head with a 12 × 46mm inset rebate, a splay-moulded face and a traditional mortar-groove in the back – which could be omitted nowadays. The 12mm edge of the doorstop has a saw kerf, or router groove, into which a Raven tubular-type silicon-rubber compression seal is push-fitted. The vertical broken lines indicate the possible division for five-fingered comb jointing. (b) This *cross section* is through a 140 × 45mm hardwood sill, with the comb-jointed doorframe jamb in the background (giving a part-elevational view of the tubular compression seal) and a partly open, weather-boarded door with a rebated bottom edge. Similar to the doorframe's rebates, the door's underside rebate is grooved and fitted with a more cylindrical-type Raven tubular silicon-rubber compression seal. This closes against the compressible fin of the upstanding PVC water-bar fitted into the grooved sill. Note that these corner-fitted, tubular-type compression seals seem to me to be better able to cope with the conflicting geometry of the closing hinge-side edge approaching the seals from a different direction to the door's top-face and lock-edge face.

The opposite edges are usually rounded or splay-moulded. Traditionally, they were bead-moulded and the head and sill were given 50 to 75mm horns on each side, which were splay-cut on site and built in to the brickwork. The former (bead-moulding) might still be used occasionally, but the latter are not. This is partly because the horns create more work for the bricklayer, partly because – if the frames are inset – the horns clash with the vertical DPC in a cavity wall,

partly because the jointing of frames seems to have changed from mortises-and-tenons (requiring horns) to comb joints (not requiring horns), and partly due to improved frame-fixing screws.

Nowadays, to improve a dwelling's energy rating in compliance with changes in Part L of The Building Regulations, more attention must be given to draught-seals and weather-seals and there is a wide (and confusing) range of these available from ironmongery companies. The sectional views illustrated above show seals in use.

## Typical door-panel mouldings

*Figures 5.7(a)(b)(c)*: Although the inner edges of door stiles and rails can of course have a square finish (which has a certain simplistic appeal) and only be grooved for panels, they are usually moulded. The three most common types are illustrated below:

## Types of door-panels

*Figures 5.8(a)(b)(c)(d)(e)*: The most common types of door-panels are 1) inset panels, embellished with stuck or planted perimeter mouldings, as at Figures 5.7(a),

(b) and (c) below; 2) raised and fielded panels, as at Figures 5.1(b) and (d) – section B-B – above (and in Chapter 3); 3) bead-and-butt panels and bead-flush panels. This final type of door panel is (or was) quite commonly used for exterior doors and is illustrated below. Such panels were used in good-quality half-glazed doors, below the middle rail and four-panelled doors, via two muntins. Basically, the panels are of 21mm par solid timber, tongued and grooved into the stiles, rails and muntins to achieve a flush appearance on the door's exterior face. The *bead-and-butt* panel is so-called because the tongued edges that abut the stiles (and the muntin) are stuck-moulded with a bead-shape mould, as illustrated at (e) below – and the top and bottom tongued edges *butt* up squarely to the horizontal rails, as illustrated at (a) and (b) below. The so-called *bead-flush* panel varies with regard to

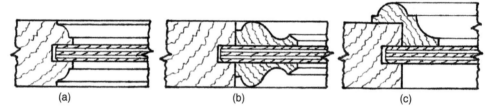

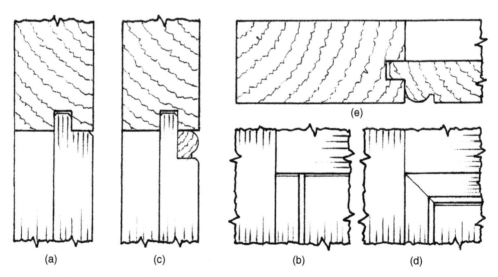

**Figure 5.7** *Part horizontal sections* through the stiles and panels of doors, showing so-called 'stuck' ovolo mouldings at (a); 'planted' Grecian ogee mouldings at (b); and a 'planted' torus/cavetto-shaped bolection moulding at (c) on one side, with a square-edged stile on the other. Note that the planted mouldings above – to allow for thermal movement between the separate components – should be diagonally pinned into the door, not the panels. However, large bolection mouldings used with raised-and-fielded panels on each side of a door, are usually slot-screwed through the panels from the interior side and the interior (covering) moulds are only pinned into the door's edges.

**Figures 5.8 (a) to (e):** These *part vertical, part horizontal sections* and the two *part elevations* show the bead-and-butt details at (a), (b) and (e) – and the bead-flush details at (c), (d) and (e). Note that bead-flush panels (showing mitred beads around all four edges) present an impractical joinery-task if the panels are stuck-moulded (as at (e) above) on their vertical, long-grain edges. This is because the top- and bottom-edges of the panels, that require cross-rebating (as at (c) above), would be obstructed for 'run-on' and 'run-off' rebating by the moulded edges that must remain (to be mitred). Stopped-rebates and chisel-pared mitres would have to be tediously made. The solution, therefore, in my view, would be to omit the stuck moulds and run a mitred, planted mould all round (as at (c) above).

its horizontal abutments to the rails. Instead of being square, the bead is returned across the grain and mitred in the corners. This is usually done – contrary to a possible shrinkage problem – by rebating the top- and bottom-edges to accommodate a planted, glued-on bead, as illustrated at (c) and (d) above.

# GLAZED DOORS

*Figure 5.9(a)(b)*: Glazed doors are related in most respects to all of the details covered here concerning panelled doors, with the obvious exception that the panels are replaced with glass – either single or double-glazed sealed units – and this will always have planted beads on one or both sides. Nowadays, the door's glazing, if positioned less than 1.5m above the floor, would have to be safety glass. Another consideration is that glazed doors with large glazed areas (especially areas with large double-glazed sealed units), should have 3mm-thick plastic 'setting-blocks' positioned under and above the units diagonally and 'locating-blocks' positioned adjacent to these on each side.

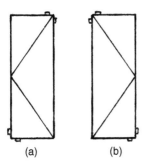

Figure 5.9 Positions of the 3mm-thick plastic setting- and location-blocks against the glass: **(a)** for left-hand side-hung doors and **(b)** for right-hand side-hung doors.

# GLAZING BARS

*Figures 5.9(c)(d)(e)*: For single-glazed internal doors, the bars are usually moulded with the traditional ovolo mould, but other designs can of course be used. A point to bear in mind with these relatively slender bars is that there are three methods of cross-jointing them and each of the methods creates a lessening of the bars' strength in one way or another. The most preferred method – and no doubt the strongest of the three – is where the *lay bars* (horizontal bars) are mortised and the vertical cross-bars are stub-tenoned and scribed to them. The three methods of jointing cross bars are shown below at (c), (d) and (e). The main reason for mortising the lay bars (and not the vertical bars) is to create a continuous, supported ledge for the glass to sit on – although there are exceptions to the rule, especially on double-hung sash windows, where the *vertical* bars would be mortised to retain continuity and give support to the slender meeting rails.

## Glazing-bar template

Figure 5.9*(f)*: Machine-planed, glazed-door components are usually ovolo-moulded and rebated after they have been through-mortised, blind-mortised, tenoned, stub-tenoned and machine-scribed, but the sequence of operations for hand-made joinery differs. After being marked out, mortised, part-tenoned (ripped down, not shouldered), ovolo-moulded and rebated, the tenons' shoulders are then cut and the ovolo scribes are made. The best technique for the hand-scribing – with a mitre template, chisel and gouge – has already been covered in Chapter 3 and Chapter 4, but using a gouge on the glazing bars is impractical. So once the outline of the ovolo scribe has been formed by chisel-paring against the template, I

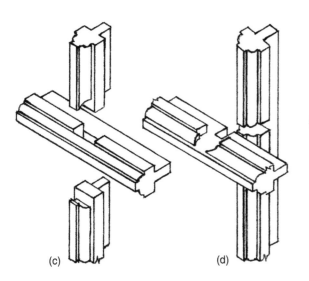

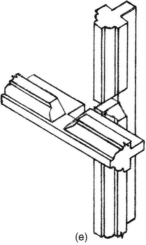

Figure 5.9: *Pictorial views* of cross-jointing to short pieces of ovolo-moulded glazing bars; **(c)** the lay bar is mortised to receive the stub-tenoned and scribed vertical bars; **(d)** the lay bar is cross-halved and scribed to receive the opposite cross-halved vertical bar; and **(e)** the lay bar is cross-halved and mitred to receive the opposite cross-halved and mitred vertical bar.

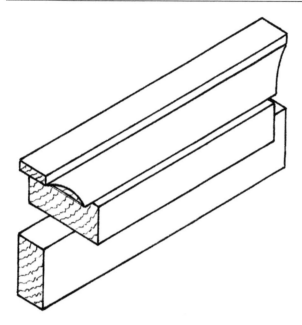

Figure 5.9 (f) *Isometric view* of a typical double-ended glazing-bar template, made from an offcut of glazing bar. If preferred, the template could be single-ended.

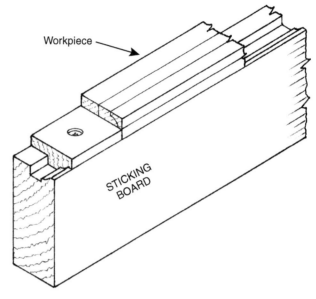

Figure 5.9 (g) *Isometric view* of the front-end of a sticking-board jig, used to make glazing bars with hand-planes or a portable powered router (by 'sticking moulds on them'; hence the origin term 'stuck moulds'). For illustrative purposes, the outline of the ovolo mould and the rebate are shown on the half-shaped glazing bar. Note the stop-end is screwed to the sticking board to allow for changing it to the opposite end, if necessary. The screw should be countersunk, as shown, to eliminate any risk of striking it when planing or routeing. My experience of using these jigs is that the inverted shape – even if not precisely made – holds the unfixed, low-rise workpiece very well.

would recommend removing the waste carefully with a coping saw (which requires a degree of practice). An alternative technique for marking out the shoulders and ovolo-mould scribes of glazing bars – especially if you are using a timber merchant's stock components – is to make a glazing-bar template from an offcut of glazing bar, as illustrated above.

## Sticking boards

*Figure 5.9(g)*: If a joiner, or a small joinery workshop, needed to make glazing bars by hand – no doubt with the aid of at least a portable powered router, as opposed to more sophisticated machinery, or (at the other extreme) ovolo- and rebate-planes – a traditional *sticking board* would be useful. This is a purpose-made jig (be it simply a certain length and thickness of board), with the inverted shape of the glazing bar formed on one of its edges. The idea is to form the rebate and ovolo mould on the edges of a board thicknessed to the bar's widest size (say 35mm), then rip this off and plane the ripped surface to the bar's narrowest size, (say 22mm). Place this, as illustrated at (g), in the inverted shape of the vice-held sticking-board jig and produce the second rebate and ovolo mould. Repeat the operation for the number of glazing bars required.

## FL&B AND L&B DOORS

As mentioned under Basic Door Knowledge, near the beginning of this chapter, FL&B and L&B are well known abbreviations in the industry for *framed, ledged, braced and matchboarded* doors – and *ledged, braced and matchboarded* doors. Note that, in non-technical parlance, the matchboarded element is not usually mentioned. This type of exterior door has been around for many decades and seems to be still popular for certain uses. The FL&B is usually seen on outbuildings other than dwellings, but the L&B is used for sheds, fence-gates and internally in period (or mock period) dwellings; they look very attractive with Norfolk thumb latches and cross-garnet T hinges. One variation in FL&B doors is that the matchboarding is sometimes tongued-and-grooved into the top, front edge of a full-thickness bottom rail (whereby the boarding becomes more like a panel), but – it seems to me – the other type that has boarding running down and fixed to the face of a bottom ledge (instead of a bottom rail) is the more common.

## Construction details of FL&B doors

*Figures 5.10(a)(b)(c)(d)(e)(f)(g)*: The ex 112 × 50mm stiles and top rail are joined with haunched mortise-and-tenon joints and the inner face-edges are chamfered, with mason's mitres where they meet in the top corners. The ex 175 × 30mm middle- and bottom-rails, reduced in thickness to allow for the boarded face – and called *ledges* instead of *rails* – are mortised and bare-face tenoned into the stiles. The bottom ledge is usually raised up by 40 to 50mm above the base of the door. The ex 112 × 30mm braces (equal to the ledge-thickness) must be diagonally positioned like gallows' brackets to support the unhinged side of the door. The braces are often right-angle pointed and fitted into the corners, but this has the potential to push the joints apart, so on good-class work they are birdsmouthed into the top rail and ledge-edges, as illustrated at (b) below. The angle-of- lean on braces above and below the middle ledge should not be less than 45° and if this cannot be achieved, they should be lined up and effectively formed as one steep angle. The ex 125 × 25mm tongued, grooved and vee-jointed (TG&V) matchboarding should theoretically cover

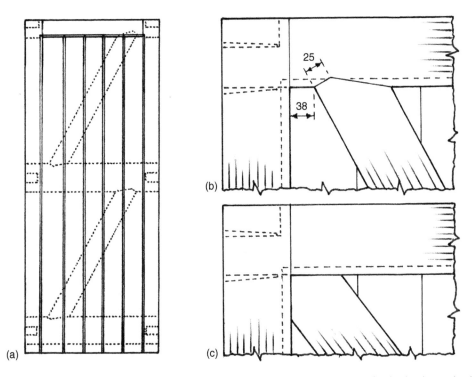

Figure 5.10 (a) *Front elevation* of a double-ledged framed, ledged and braced door; (b) the birdsmouthed housing-detail at the top of a diagonal brace; and (c) the inferior method of a right-angle pointed brace positioned in the corner. Note: in practice, the most successful way of achieving a good birdsmouth fit to the top rail and ledges is to lay the un-jointed braces in their exact position on the framed-up door, mark the inner door-edges where the braces cross and, immediately above these points, mark the braces' edges. Use these marks to form the obtuse-pointed ends of the braces, then lay them back (in their numbered positions) and mark the birdsmouth shapes with a sharp pencil. When forming the birdsmouth shapes by saw cuts and chisel-paring, leave the pencil lines showing to achieve a snug fit.

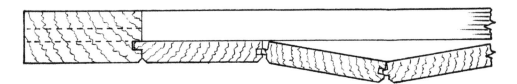

Figure 5.10 (d) Part *horizontal section* through a FL&B door, showing the final two TG&V boards to be sprung in against the ledges – and the broken lines indicating the position of the bare-faced tenon. Note that the left-hand and right-hand edge-boards (and the top edges of all the boards) are usually given a thicker tongue (about one-third of the boards' thickness), on the inner edges only, as shown above and below at (g).

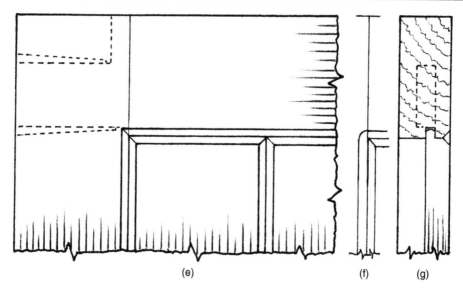

(e)          (f)   (g)

Figure 5.10 (e) Part *elevation* of a FL&B door, showing the hidden detail of the haunched and wedged mortise-and-tenon joint, the vee-jointed boards with a return-chamfer on their top edges and the chamfered edges of the top rail and stile joined by a mason's mitre; (f) shows the appearance of a mason's mitre formed with a portable powered router; and (g) shows a part *vertical section* through the top rail and TG&V boarding, exposing its bare-faced tongue into the rail, its chamfered top-edge and indicating the position of the M&T joint.

the face-side in equal widths – otherwise, any reduction should be made to one of the boards on each side.

On assembly, after the frame and ledges have been glued, cramped and wedged, the preservative-treated, unglued T&G boards should be left at least 12mm down from the underside of the top rail, sprung-in to a tight fit in width and then knocked up carefully to fit the top rail's groove. Traditionally, the boards were *clench-nailed* with 38mm *cut-clasp* nails, two per fixing, in a staggered pattern. After being driven under the surface by 2 to 3mm with a nail-punch, they were clenched over on the ledge-side, in the direction of the grain and punched under the surface again. This seemingly simple final operation requires a degree of developed skill, otherwise the fixings can be loosened and the heads of the nails brought to the surface or left protruding on the face-side of the door. The recommended technique is to bend the nails over to about 60°, then drive them sideways and under the surface with a large-headed nail-punch held initially at about 30° against the nails' lowest point to the timber's surface. Although cut-clasp nails are still available, they appear not to be used on these types of doors anymore, yet I rate their clenched holding-power as being well above any alternative nail or other fixings used in recent years.

## Construction details of L&B doors

*Figures 5.11(a)(b)(c)(d)(e)(f)*: The making of ledged, braced and matchboarded doors is relatively straight-forward. The TG&V-jointed boards are simply

cramped together face-down with at least three sash cramps, then a top ledge of ex 100 × 30mm and a middle- and bottom-ledge of ex 150 × 30mm are laid across the boards in their pre-marked positions and screwed at their ends, as illustrated. The birdsmouthed brace-joints are marked and formed (as already

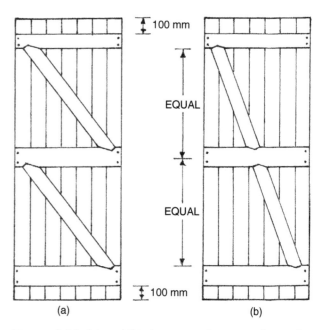

(a)          (b)

Figures 5.11 (a) and (b) show *rear elevations* of typical ledged, braced and matchboarded doors with (a) birdsmouthed braces at 52° above and below the middle ledge; and (b) birdsmouthed braces in one line at 70° above and below the ledge.

described), then the cramped assembly is carefully turned over and the boards are clench-nailed to the ledges in a staggered nailing pattern. Next, the braces are fitted and are also fixed with clenched cut-clasp nails, as indicated at (c) and (d) below.

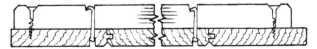

**Figure 5.11 (c)** Part *horizontal section* through an L&B door, showing clench-nailing into a ledge-rail and alternative screwed-end methods: The ledge-ends on the left side are only set back by 4mm for 'shooting-in' the door on site. This method should always be used if the Tee hinges are to be fixed on the face of the ledges and can also be used if the hinges are to be fixed on the boarded face, except that the hinge-side door-stop would then require notching out to accommodate the three ledges. The ledges on the right-hand side are set back by 16mm; this is 4mm for shooting-in and 12mm for an un-notched door-stop. The left-hand method at each end would be stronger.

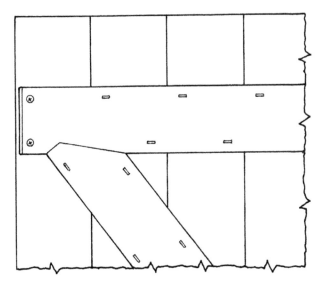

**Figure 5.11 (d)** Part *rear elevation* of L&B door, showing the screw fixings and the staggered clench-nailing bent over in the grain-direction of the ledge and brace.

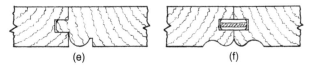

Figures 5.11 (e) and (f) are Part *horizontal section views* of alternative jointing for L&B doors: (e) is T&G bead jointing; and (f) is T&G lamb's tongue jointing. The 3-ply tongue (which can be of hardwood) must be glued in one groove only.

# FLUSH DOORS

Nowadays, because of sophisticated and cost-effective manufacturing techniques, flush doors are not usually made by individual joiners and joinery workshops unless a customer wants an odd non-standard-size door made to measure. The problem is the absence of an industrial-type press for cramping the outer skins of plywood (or hardboard) to the glued skeletal frame of the door – and having to resort to using contact adhesives with a *grab* capability, requiring only hand-pressure (which should be applied with a semi-hard rubber veneer roller, or a veneer hammer). Apart from Bostik's Evo-Stik Impact Instant-Contact Adhesive, which I have used for making flush doors for fitted wardrobes, there are a number of modern grab-adhesives that might do the job for interior- *and* exterior-door skins. One is called *Sticks Like Sh-t* (*Serious Stuff* is its dual name), another, just marketed, is *Serious Glue Liquid*, both types by Bostik. Another exterior type worth considering is Apollo's A26 General Purpose Contact Adhesive.

## Skeletal framework

*Figure 5.12(a)(b)*: For purpose-made flush doors, a lightweight frame of, say 34 × 28mm par, is used for the stiles, top and bottom rails and the intermediate rails. Usually a lock-block is integrated on one or both sides, as shown in Figure 5.12(a). The jointing of the framing consists of a continuous 10 × 10mm groove on the inner edges of the stiles with the short stub-tenoned ends of all the rails glued into them. The rails, as illustrated at (b), should have a 4 × 4mm notch or a 5mm diameter hole made in them to inhibit trapped air causing distortion to the door-skins and to allow the trapped air to equalize with the external air. The hanging-edge and the lock-edge of the door are lipped with 10mm-thick lipping – which is glued and cramped on after the slightly-overhanging ply- or hardboard-skin faces have been cleaned off.

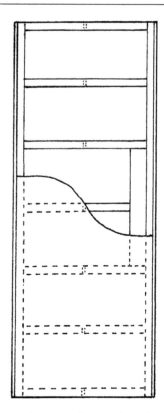

**Figure 5.12 (a)** *Elevation* of internal, partly clad plywood or hardboard flush door.

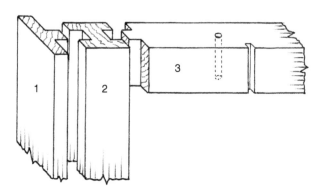

**Figure 5.12 (b)** *3D view* of parts of the lightweight framing: **1)** *Tongued lipping* of ex 42 × 20mm finish (2mm added to width for 1mm cleaning-up on each door-face); note, the lipping can also be of a 42 × 10mm finish, without a tongue. **2)** *Edge-framing stiles* of 34 × 28mm finish with 10 × 10mm grooves; and **3)** *Rails* of 34 × 28mm finish, stub-tenoned and drilled or notched – (drilling is preferable).

# MAKING AN UPGRADED EXTERIOR DOOR

*Figures 5.13(a)(b)(c)(d)(e)*: Although there is a wide range of modern and traditional designs for exterior timber doors (on view to anyone out walking), I have designed a half-glazed door with a revamped traditional panel-type that has an established reputation for durability and longevity. Essentially, it has a *bead-flush* panel and, in my design, it could also be a *bead-and-butt* panel (illustrated and described at Figures 5.8(c) and (d) above). As seen in the section views below, I have upgraded the door with a layer of 25mm-thick Celotex rigid insulation board sandwiched between the 9mm-thick outer panels of exterior-grade plywood. Additionally, the glazed area above has a 16mm sealed unit with low-E (low emissive) glass and warm-edge spacer bars – and weather seals are an integral part of the doorframe.

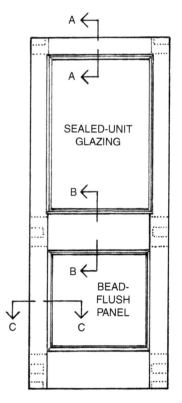

**Figure 5.13 (a)** *Elevation* of an upgraded half-glazed exterior-type door with a revamped traditional panel below and a 16mm double-glazed sealed unit above.

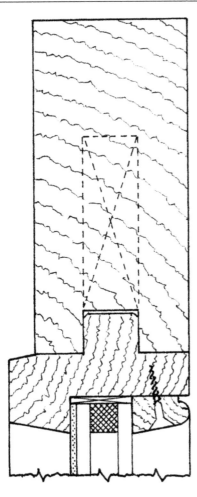

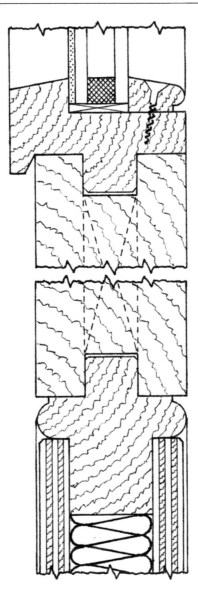

**Figure 5.13 (b)** *Section A-A* through the 95 × 45mm par finished top rail showing the outline of the 16mm concealed mortise-and-tenon position and its related size; the ex 60 × 38mm tongued-and-rebated aperture-lining necessary to convert the stile's panel-groove to an acceptable-sized rebate for sealed-unit glazing; the 16mm sealed unit bedded in butyl glazing compound; and the inner, splayed glazing bead (beaded to match the bottom panels) screwed – or cup-screwed for future replacement of the sealed unit.

**Figure 5.13 (c)** *Section B-B* through the centrally broken depth of the 195 × 45mm par finished middle rail, grooved for the aperture-lining above and the purpose-made panel below. The two concealed mortise-and-tenon positions (with a middle haunch) must be visualized from the incomplete broken-lines; the purpose-made panel consists of ex 50 × 50mm cross-shaped, tongued, rebated and twice-beaded framing, mitred and glued at its corners, with 25mm Celotex rigid insulation sandwiched between two 9mm-thick exterior grade plywood panels, as illustrated. The panel, of course, is snugly fitted without being glued. The sealed unit above is shown on setting blocks.

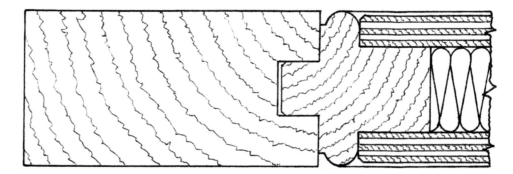

Figure 5.13 (d) *Section C-C* through the 95 × 45mm par finished stile, showing similar detailed features to those described in Section B-B.

## Construction details

*Figure 5.13(e)*: Note that although the made-up panel obviously must be fitted into position at the assembly of the door and gluing-up stage, because the sill and head of the aperture linings are best run through on each side, with the side-linings scribed onto them, they should be fitted and glued in after the door has been assembled.

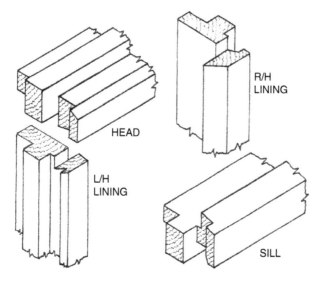

Figure 5.13 (e) *Isometric views* of four of the scribed ends of the aperture-lining. Note that a template (similar to that shown at Figure 5.9(f) above) could be made for these.

# 6

# Designing and making stairs to current building regulations

## INTRODUCTION

Means of climbing up or down from one floor level to another in the form of steps or stairs, have been in existence for many centuries. However, in the majority of the UK's London houses in the 13th century, stairs were only crude arrangements of upright poles with projecting pegs, unsafe ladders and barked tree trunks roughly notched into triangular-shaped steps. Stairs only really developed in the 17th century, after the Great Fire of London. This historic event brought about the first comprehensive Building Act in 1667.

Since then, many Building and Public Health acts have been in force and numerous building byelaws have been operating in different parts of the country – often in contradiction to each other. Not until 1965 were The Building Regulations introduced to replace the byelaws and establish uniformity.

These Regulations, having been amended and revised many times since, are now embodied in separate publications known as *Approved Documents* (often abbreviated to AD). In this chapter, we are only concerned with AD K1 which controls the design, construction and installation of *Stairways, ramps and ladders* in England and Wales – but not Scotland or premises which are occupied by the Crown. In practice, however, whether controlled or not, good stair-design nowadays cannot deviate from certain established principles.

## DETAILS OF DESIGN

*Figures 6.1(a)(b)*: The first thing to understand in the basic theory of stair design is that one movement of a person's foot *going* forward is referred to as '*the going*' of the step and the other foot *rising* up (to another level) is referred to as '*the rise*' of the step. Therefore, from a design point-of-view, all of the step-movements going forward are referred to as '*the total going*' *(TG)* and all of the step-movements rising up are referred to as '*the total rise*' *(TR)*. These two terms (TG and TR) are important references to the simple

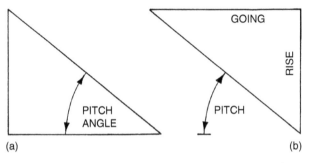

**Figures 6.1 (a)** and **(b)** The concept of a basic triangle transposed into a step.

maths and geometrical division required to design legal flights of stairs. Mathematically, a step can be related to a right-angled triangle, whereby its base-line represents *the going* of a single step, the adjacent side represents *the rise* – and the ratio of each to the other, forming an hypotenuse, determines the legally-important pitch angle of a step, as illustrated at Figures 6.1(a) and (b).

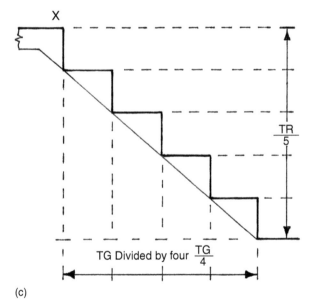

**Figure 6.1 (c)** A multiple of steps related to the division of the total going and total rise.

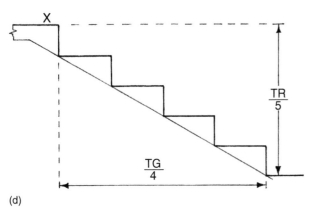

(d)

**Figure 6.1 (d)** Although the scaled total-rise and the number of divisions at 6.1(c) is equal to the illustration at 6.1(d), it can be seen how an increased going of each step alters the pitch angle. Figure 6.1(c) has a 40° pitch; 6.1(d) has 29°.

*Figure 6.1(c)*: The above concept of relating a step to a triangle can also be applied to a multiple of steps, as illustrated above, whereby the 'total going' of all the steps (equally divided horizontally between the face of the top step and the face of the bottom step) and – likewise – the 'total rise' of all the steps (equally divided between the lower- and the upper-floor levels), forms a pitch angle according to the ratio of the total going (TG) related to the total rise (TR).

*Figure 6.1(d)*: In considering an acceptable pitch-angle, the total rise is of course a constant, but as shown by comparing Figures 6.1(c) and 6.1(d), an adjustment to the total going alters the angle of pitch. This is an important factor in stair design, especially when trying to fit a staircase into a small space where there is a limited TG.

As seen in Figures 6.1(c) and (d) above, TR (total rise) is divided by five, unlike TG (total going) which is divided by four. This is because the top steps – marked X in each Figure – are in fact part of the landing and therefore are not included in the horizontal division. As the top surface of a step is called a 'tread' and the vertical face is called a 'riser', as a rule it can be said that in any single flight of stairs, *there must always be one less tread than risers*. When initially designing a staircase and calculating how many steps can be used to meet the various regulations, this basic rule ensures a correct trial-and-error division of TR related to TG.

Although straight flights of stairs are quite common, they are not universally regarded as being very attractive and they are not always possible with a restricted going, i.e. a limited stairwell length. For these reasons, changes in the direction of the stair are often made.

## DIRECTIONAL CHANGES

If you think of directional changes – in any stair design – as they would relate to a spiral staircase having a complete turn of 360°, then it can be readily seen that a *quarter-turn stair* changes direction by 90° and a *half-turn stair* changes direction by 180°. Although *three-quarter-turn stairs*, changing direction by 270°, can be found in older-style properties, they are not seen nowadays in modern dwelling houses. Apart from straight flights, quarter-turn and half-turn are the two most common types of turning stairs – and are turned by introducing either a quarter- or half-turn landing (an intermediary platform), or tapered (winding) steps. With tapered steps (the modern term), because they are considered to be potentially more dangerous, it is safer to keep them at or near the bottom of a staircase, if possible.

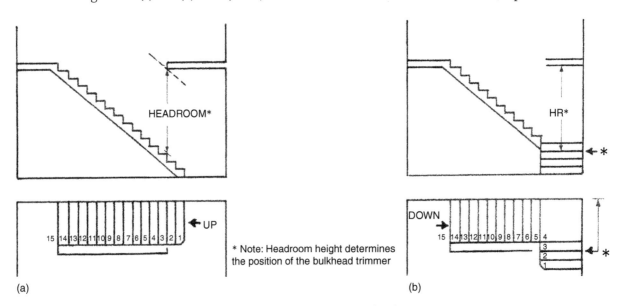

(a)

* Note: Headroom height determines the position of the bulkhead trimmer

(b)

**Figure 6.2 (a)** Straight-flight stair; **(b)** Quarter-turn stair via a quarter-turn landing.

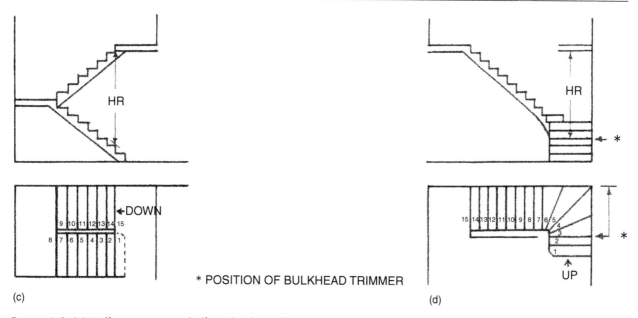

**Figure 6.2** (c) Half-turn stair via a half-turn landing (d) Quarter-turn stair via four tapered steps. Note that – at the scaled-design stage – the minimum legal headroom has been worked out to be above the third riser (*).

*Figures 6.2(a)(b)(c)(d)*: These plan- and elevation-diagrams outline the four most common stair arrangements in use. Although variations to these basic designs can be made, they serve to compare the effect on available floor space – and show where the 'headroom' (HR) regulation applies – and highlights the application of the headroom measurement for positioning the trimmed stairwell-opening.

## ADDITIONAL CHANGE

*Figure 6.3*: In the diagrams in Figures 6.2(a), (b) and (d), another change in direction could be achieved by making allowance for a quarter-turn landing at the top of each flight. This would reduce the total going by one step – if required – as indicated in Figure 6.3 and

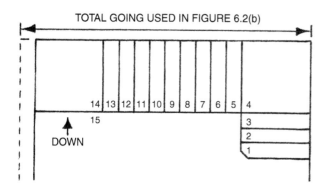

**Figure 6.3** A quarter-turn landing added at the top reduces the total going by one step and changes the designation to a half-turn stair.

compared with Figure 6.2(b). Although, a single step positioned at the head of a stair can be dangerous, as it is not always noticed.

## LIMITING FACTORS

There are many limiting factors to contend with in stair design and in site-access for the assembled flights. These include: 1) clashing door- and/or window-positions; 2) structural intrusion of piers and offset walls; 3) serving split (mezzanine) floors with inconsistent storey heights; 4) striking a regulatory balance between shallow and steep stair pitches; both of which are tiring to ascend and the latter type being dangerous to descend; 5) achieving sufficient regulatory headroom – defined in Figures 6.2(a) and (c) – so as not to cause a tall person to stoop or suffer head injury; and 6) achieving good access and adequate 'clearance' for the movement and manoeuvrability of furniture, etc, up and down stairs. Where access to a stairwell is restricted and/or awkward – which is not unusual when replacement or additional stairs are being installed in a property – the makers and fixers of the stairs often have to complete the assembly of the separate made-up steps and the strings by improvisation on site.

Stair design, then, is essentially an integral part of the whole design of a building and perhaps only concerns the architect – although to implement the architect's design from small-scaled drawings (or take on the role as designer and maker) means we need at least as much technical knowledge as the designer. Also, we need a good understanding of the Building Regulations.

# STAIR REGULATIONS GUIDE

Approved Document K1 (AD K1) of the current Building Regulations (amended in 2010, but with no substantive changes to the 1998 Edition and 2000 amendments – although further amendments are due) governs the design of internal and external stairways and balustrades, etc, in buildings and differentiates between three categories: *1) private stairs*, intended to be used for only one dwelling; *2) institutional and assembly stairs*, serving a place where a substantial number of people will gather; and *3) other stairs*, for all other buildings apart from categories 1 and 2.

The following modified version of AD K1 covers most of the points concerning stairs and balustrades only – and, by use of additional figured illustrations and text in places, an attempt has been made to present a clearer picture of stair regulations as a guide, but not as a substitute.

## Definitions

The following illustrated definitions are given to terms used in AD K1 and a few others have been added for greater clarity. As these are only basic definitions, more explanatory information will be developed progressively throughout the chapter.

## Alternating tread stair

*Figure 6.4*: This is a stair constructed of paddle-shaped treads with the wide portion alternating from one side to the other on consecutive treads.

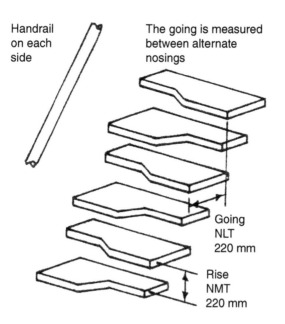

Figure 6.4 A *three-dimensional impression* of an Alternating Tread Stair.

## Balustrade

*Figure 6.5*: A balustrade is a protective barrier at the side of a stair, landing or balcony, etc, usually comprised of newel posts, handrails and balusters (sometimes referred to as 'spindles') on wooden stairs, but which may also be panelling, a wall, parapet, screen or railing, etc.

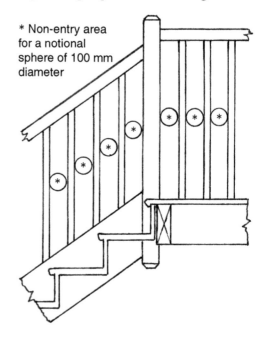

Figure 6.5 Protective balustrade on a stair and landing, showing regulatory reference (*) to the spaces between the balusters.

## Deemed length

*Figure 6.6*: If consecutive tapered treads are of different lengths, as illustrated below (when they extend into a corner), each tread can be deemed to have a length equal to the shortest length of such treads. Although not now referred to in AD K1, the deemed length (DL) needs to

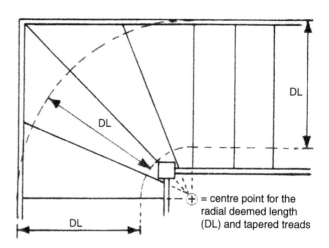

Figure 6.6 Application of Deemed Length (DL) for reference to other Regulations.

be established on certain stairs (see Figures 6.23(a)(b) (c)) for the purpose of defining the extremities to which the pitch line(s) will apply for checking the 2R+G rule.

## Flight

The part of a stair or a ramp between landings, constructed with a continuous series of steps or a continuous slope.

## Going

*Figure 6.7*: The horizontal dimension from the nosing edge of one tread to the nosing edge of the next consecutive tread above it, as illustrated.

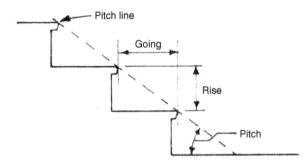

Figure 6.7 The Going, Rise, Pitch and Pitch line.

## Rise

*Figure 6.7*: The vertical dimension of one unit of the total vertical division (total rise: TR) of a flight of stairs, as illustrated.

## Pitch

*Figure 6.7*: This refers to the degree of incline from the horizontal to the inclined pitch angle of the stair.

## Going of a landing

*Figures 6.8 and 6.9*: This is the horizontal dimension determining the width of a landing measured

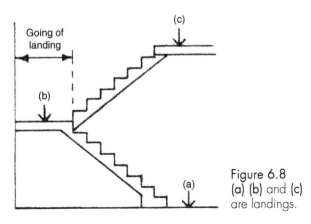

Figure 6.8 (a) (b) and (c) are landings.

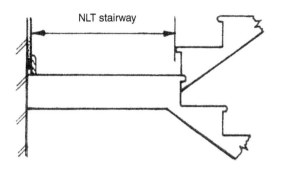

Figure 6.9 The going of landings to be NLT (not less than) the stairway width. Note that it is not clear in K1 whether this definition is from the face of a wall, or from the face of the skirting board. Designers might be wise, therefore, to assume the latter.

at right angles to the top- or bottom-step, from the nosing's edge to the opposite wall surface or balustrade.

## Nosing

*Figure 6.10*: The nosing is the projecting front edge of a tread (of whatever material) past the face of the riser. In traditional joinery terms, when the tread is of timber, the rule is that the projection should not be more than the tread-board's thickness.

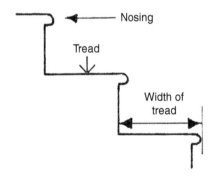

Figure 6.10 Nosings related to risers and treads.

## Stair

A stair is a succession of steps and landings that makes it possible to pass on foot to other levels.

## Tapered tread

*Figures 6.6 and 6.11*: This is a step in which the nosing is not parallel to the nosing of the step or landing above it. Traditionally, these treads were referred to as 'winders' or 'winding steps' until the Building Regulations renamed them.

## Pitch line

*Figures 6.7 and 6.11*: This is a notional (imaginary) line used for reference to the various rules, which connects the nosings of all the treads in a flight and also serves as a line of reference (in a midway, central position) for measuring the regulatory stair-formula known as '2R+G' (see Figure 6.22) on tapered tread steps.

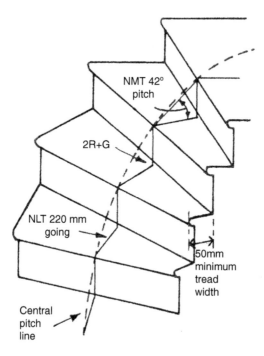

Figure 6.11 Pitch line related radially on tapered steps; and the minimum 'tread-width' of 50mm – somewhat relaxed from a previous K1 reference to the minimum 'going' being NLT 50mm. Note the abbreviations NLT = not less than, and NMT = not more than.

## Helical stair

A stair that describes a helix around a central void (with an open stairwell). Such a stair is traditionally known as a geometrical stair.

## Spiral Stair

A stair that describes a helix around a central supporting column (*without* an open stairwell).

## General requirements for stairs

### Steepness of stairs

In a flight, the steps should all have the same rise and the same going to the dimensions given here further on for each category of stair in relation to the 2R+G formula.

## Alternative approach

AD K1 states that the requirement for steepness of stairs can also be met by following the relevant recommendations in BS 5395 *Stairs, ladders and walkways*: Part 1: 1977 *Code of practice for the design of straight stairs*.

## Level steps and open risers

*Figure 6.12*: Steps should have level treads and may have open risers, but treads should then overlap each other by at least 16mm. For steps in buildings providing the means of access for disabled people, reference should be made to *Approved Document M (AD M): Access and facilities for disabled people*.

All stairs that have open risers and are likely to be used by children under 5 years of age should be constructed so that a notional sphere of 100mm diameter could not pass through the open risers.

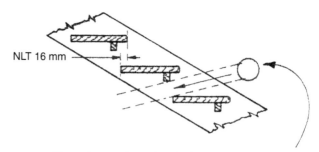

Non-entry area for notional sphere of 100 mm diameter

Figure 6.12 Open-riser stair restrictions.

## Headroom

*Figures 6.13 and 6.14*: Clear headroom of NLT 2m measured vertically from the pitch line to the soffit of the stair above, or to the corner-edge of the bulk-head of the landing, etc, must be achieved. For loft

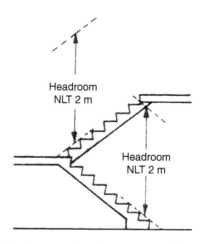

Figure 6.13 Minimum Headroom required from notional pitch line of NLT (not less than) 2m.

conversions where there is not enough space to establish this height, the headroom will be satisfactory if the height at the centre of the stair-width is 1.9m, reducing to 1.8m at the side of the stair, as illustrated at 6.14.

## Clearance

Clearance, not now referred to in the amended AD K1, was a traditional requirement of NLT 1.5m between flights, measured at right angles to the pitch line of a stair, to the underside of any flight or structural intrusion above.

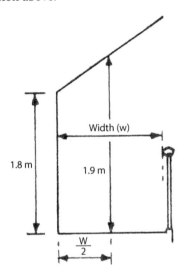

**Figure 6.14** Reduced Headroom, if required, for loft conversions.

## Width of flights

*Figure 6.15*: Contrary to previous regulations (which gave 800mm as the minimum unobstructed width for the main stair in a private dwelling), no recommendations for minimum stair widths are now given. However, designers should bear in mind the requirements for stairs which:

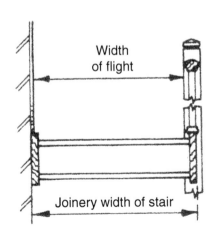

**Figure 6.15** Width of flight.

- form part of means of escape (reference should be made to *Approved Document B: Fire Safety*;
- provide access for disabled people (reference should be made to *Approved Document M: Access and facilities for disabled people*).

## Dividing flights

*Figure 6.16*: A stair in a public building which is wider than 1.8m should be divided into flights with relief handrails which are not more than 1.8m, as illustrated.

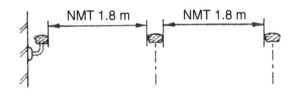

**Figure 6.16** Division of flights over 1.8m wide.

## Length of flights

*Figures 6.17 and 6.18*: The number of risers in a flight should be limited to 16 if a stair serves an area used as a shop or for assembly purposes. Stairs having more than 36 risers in consecutive flights should have at

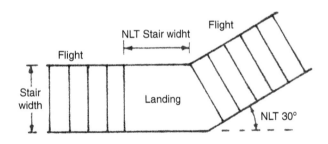

**Figure 6.17** Change of direction via a landing.

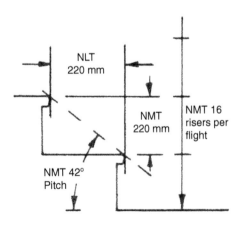

**Figure 6.18** Category 1 Stair, indicating the maximum permissible pitch, rise and risers per flight – and the minimum going.

least one change of direction between flights of at least 30° in plan, as illustrated.

## Landings

*Figures 6.8, 6.9, 6.19(a) and (b)*: A landing should be provided at the top and bottom of every flight. The width and length of every landing should be at least as long as the smallest width of the flight and may include part of the floor of the building. To afford safe passage, landings should be clear of permanent obstruction. A door may swing across a landing at the bottom of a flight (Figure 6.19(a)), but only if it will leave a clear space of at least 400mm across the full width of the flight. Doors to cupboards and ducts may open in a similar way over a landing at the top of a flight (Figure 6.19(b)). For means-of-escape requirements, reference should be made to *Approved Document B: Fire Safety*. Landings should be level unless they are formed by the ground at the top or bottom of a flight. The maximum slope of this type of landing may be 1 in 20, provided that the ground is paved or otherwise made firm.

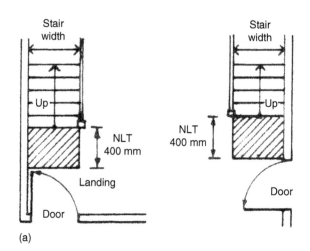

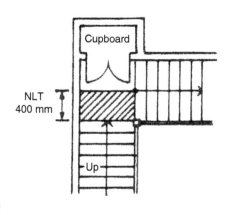

Figure 6.19 (a) Landings next to doors; and (b) a cupboard on a landing.

## Rise-and-Going limits for each category of stair

### Category 1 – Private Stair

*Figure 6.18*: Any rise between 155mm and 220mm can be used with any going between 245mm and 260mm, or any rise between 165mm and 200mm can be used with any going between 220mm and 300mm.

### Category 2 – Institutional and Assembly Stair

*Figure 6.20*: Any rise between 135mm and 180mm can be used with any going between 280mm and 340mm. Note that for maximum rise of stairs providing means of access for disabled people, references should be made to *Approved Document M: Access and facilities for disabled people*.

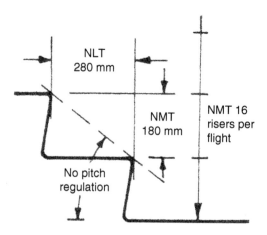

Figure 6.20 Category 2 Stair, indicating the maximum rise and risers per flight, the minimum going and an unregulated pitch.

### Category 3 – Other Stair

*Figure 6.21*: Any rise between 150mm and 190mm can be used with any going between 250mm and 320mm. Note that reference to *Approved Document M: Access and facilities for disabled people* also applies here.

### Spiral and helical stairs

It is further recommended in AD K1 that stairs designed in accordance with BS 5395: *Stairs, ladders and walkways* Part 2: 1984 *Code of practice for the design of helical and spiral stairs*, will be adequate. Stairs with goings less than shown in this standard may be considered in conversion work when space is limited and the stair does not serve more than one habitable room.

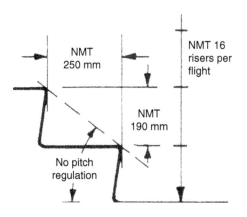

Figure 6.21 Category 3 Stair, indicating the maximum rise and risers per flight, the minimum going and – again – an unregulated pitch.

## Pitch

*Figure 6.18*: The maximum pitch for a Private Stair is 42°. Note that the recommended pitch angles for the other two categories of stair are not given in AD K1. However, using the criteria that are given, if the maximum rise and the minimum going were used in these categories, the maximum possible pitch for Category 2 would be 33° and for Category 3 would be 38°.

Note that if the area of a floor of a building in Category 2 (Institutional and Assembly Stair) is less than 100m², the going of 280mm may be reduced to 250mm. Therefore, with the maximum rise and the minimum going, the maximum possible pitch would be increased from 33° to 36°.

## The 2R+G design formula

*Figure 6.22*: In all three categories, the sum of the *going* plus twice the *rise* of a step (traditionally established and commonly referred to as *2R+G*) should be *not less than* (NLT) *550mm, nor more than* (NMT) *700mm* (subject to the separate criteria laid down for tapered treads).

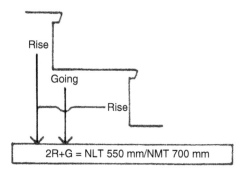

Figure 6.22 Graphic illustration of Stair-Design Formula.

## Special stairs

### Tapered treads

*Figures 6.11 and 6.23(a)(b)(c)*: For a stair with tapered treads, 2R+G should be measured as follows:

- If the width of the stair is less than 1m, 2R+G should be measured tangentially on the centre-line, where the notional curved pitch line touches the nosings (Figures 6.11 and 6.23(a)).
- If the width of the stair is 1m or more, 2R+G should be measured at 270mm from each side, tangentially where each curved pitch line touches the nosings (Figures 6.23(b) and (c)). The minimum 'tread-width' of tapered treads, from the face of a riser to the outer-edge of the adjacent tread, should not be less than 50mm (as shown in Figure 6.11). Where consecutive tapered treads are used, a uniform going should be

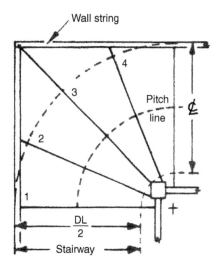

Figure 6.23 (a) Deemed length (DL) divided by two to give a single pitch line (for measuring 2R+G) if the stairway is *less than 1m* wide.

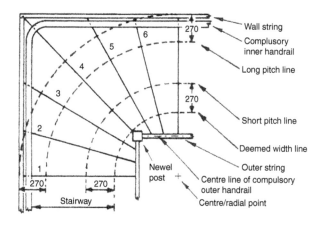

Figure 6.23 (b) Two pitch lines required for measuring the minimum and the maximum 2R+G for stairways of *1m or more* in width (deemed length).

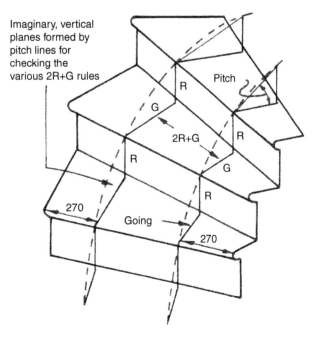

Imaginary, vertical planes formed by pitch lines for checking the various 2R+G rules

**Figure 6.23 (c)** Imaginary, triangular vertical planes shown to emphasise the measuring points for minimum 2R+G (R/H pitch-line) and maximum 2R+G (L/H pitch line) on tapered-tread stairways of *1m or more* in width (deemed length).

maintained. Where a stair consists of straight *and* tapered treads, the going of the tapered treads should not be less than the going of the straight flight.

Note that BS 585: *Wood Stairs* Part 1: 1989: *Specification for stairs with closed risers for domestic use, including straight winder* (tapered tread) *flights and quarter- or half-space* (half-turn) *landings* is given in AD K1 as a British Standard which will offer reasonable safety in the design of stairs.

## Alternating tread stair

*Figure 6.4*: This type of stair is designed to save space and has alternate handed steps with part of the tread cut away; the user relies on familiarity from regular use for reasonable safety. Alternating tread stairs should only be installed in one or more straight flights for a loft conversion and then only when there is not enough space to accommodate a stair which satisfies the criteria already covered for *Private Stairs*. An alternating tread stair should only be used for access to one habitable room, together with, if desired, a bathroom and/or a WC. The WC must not be the only one in the dwelling. Steps should be uniform with parallel nosings. The stair should have handrails on both sides and the treads should have slip-resistant surfaces. The tread sizes over the wider part of the step should have a maximum rise of 220mm and a minimum going of 220mm and

should be constructed so that a 100mm diameter notional sphere could not pass through the open risers.

## Fixed ladders

A fixed ladder should have fixed handrails on both sides and should only be installed for access in a loft conversion – and then only when there is not enough space without alteration to the existing space to accommodate a stair which satisfies the criteria already covered for *Private Stairs*. It should be used for access to only one habitable room. Retractable ladders are not acceptable for means of escape. For reference to this, see *Approved Document B: Fire safety.*

## Handrails for stairs

*Figures 6.24(a) and (b)*: Stairs should have a handrail on at least one side if they are less than 1m wide and should have a handrail on both sides if they are wider. Handrails should be provided beside the two bottom steps in public buildings (Category 2 Stairs) and where stairs are intended to be used by people with disabilities. In other places (Category 1 Stairs), handrails need not be provided beside the two bottom steps – as indicated in Figure 6.24(b).

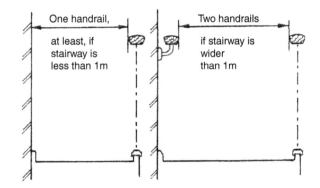

**Figure 6.24 (a)** Handrail rules for stairs.

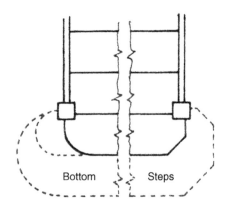

**Figure 6.24 (b)** Steps allowed in Category 1 Stairs without handrails.

## Regulatory handrail heights

*Figure 6.25*: In all buildings, handrail heights should be between 900mm and 1000mm (1m), measured to the top of the handrail from the notional pitch line or floor. Handrails can form the top of a guarding, if the heights can be matched; (the *matching* is usually effected at the transitional point of the upper-flight newel post – where the inclined handrail can be seen to be lower than the level/landing handrail – Figures 6.25 and 6.26.

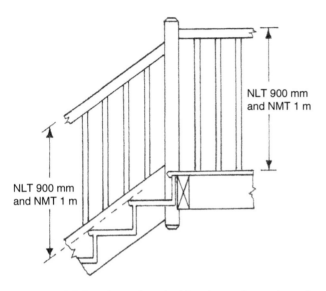

**NLT 900 mm and NMT 1 m**

**NLT 900 mm and NMT 1 m**

**Figure 6.25** Regulatory handrail heights and guarding of Stairs.

## Guarding of stairs

*Figures 6.5, 6.24, 6.25 and 6.26*: As illustrated in these stair notes, flights and landings should be guarded at the sides:

● In dwellings when there is a drop of more than 600mm;
● In other buildings when there are two or more risers.

The guarding to a flight should prevent children being held fast by the guarding, except on stairs in a building which is not likely to be used by children under 5 years. In the first case, the construction should be such that:

● A 100mm diameter notional sphere cannot pass through any openings in the guarding (Figure 6.5);
● Children will not readily be able to climb the guarding (this in effect outlaws the horizontal ranch-style balustrades of the 1970s (Figure 6.26). However, amended Building Regulations are not applied retrospectively – although changing such balustrades is a fairly straightforward carpentry operation.

The height of the guarding (balustrade) is given in Figure 6.25 for single-family dwellings in Category 1. External balconies should have a guarding/balustrade of 1100mm (1.1m).

## Staircase

Although the term *staircase* is often used instead of the term *stair*, to be technically correct, *staircase* refers to the complete structure and includes the stair, balustrades, additional newel posts, if any, apron linings (that cover the faces of trimmer and trimming joists forming the stairwell opening) and any spandrel framing, etc.

## Spandrel

The triangular panelling, studded wall or cupboard framing under a flight of steps (stair), usually positioned directly under the outer string.

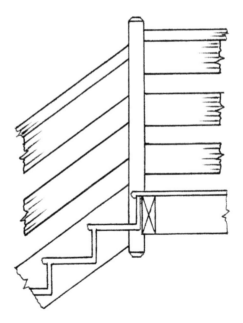

**Figure 6.26** As inferred in part 1.29b of AD K1, Ranch-style balustrades as above – or similar constructions that might be readily climbable by children – are not permitted.

## Meeting the various regulations

*Figure 6.27*: This illustration shows a 1:10 scaled working drawing of a half-turn stair for a loft-conversion. And although winding steps – as previously mentioned – are not ideal at the head of a stair, six tapered treads are used because of the restricted total going. The stair was designed by me and taken on by Hastings College as an exercise. Two of my

Advanced Craft joinery students made it and helped me to install it in 1989. Note the listing of the various regulations that are ticked as having been met. Meticulous checks like this are essential at the design stage.

# SITE MEASUREMENTS

Traditionally, accurate information required for the production of wooden stairs was considered essential; measurements on working drawings produced by others (and the newly-built stairwell) were not assumed to be correct, unless checked on site by an experienced, responsible person.

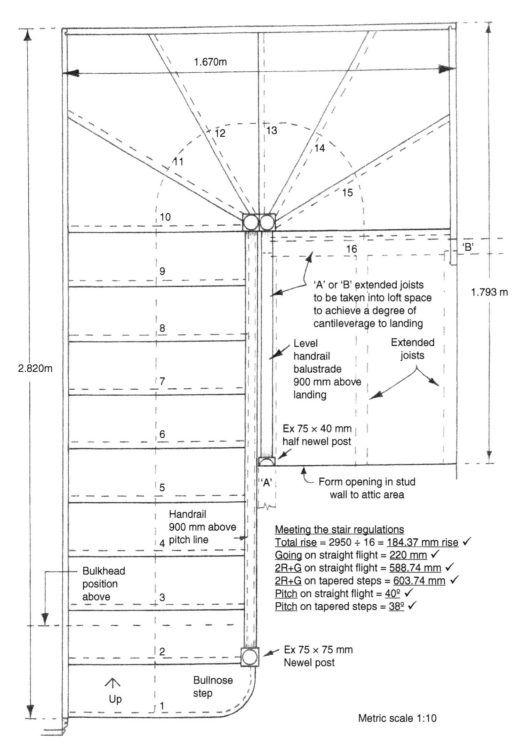

1.670m

2.820m

1.793 m

'A' or 'B' extended joists to be taken into loft space to achieve a degree of cantileverage to landing

Level handrail balustrade 900 mm above landing

Extended joists

Ex 75 × 40 mm half newel post

Form opening in stud wall to attic area

Handrail 900 mm above pitch line

Bulkhead position above

Meeting the stair regulations
Total rise = 2950 ÷ 16 = <u>184.37 mm rise</u> ✓
<u>Going</u> on straight flight = <u>220 mm</u> ✓
<u>2R+G</u> on straight flight = <u>588.74 mm</u> ✓
<u>2R+G</u> on tapered steps = <u>603.74 mm</u> ✓
<u>Pitch</u> on straight flight = <u>40º</u> ✓
<u>Pitch</u> on tapered steps = <u>38º</u> ✓

Ex 75 × 75 mm Newel post

Bullnose step

↑
Up

Metric scale 1:10

Figure 6.27
Staircase designed and installed in an actual loft conversion.

Nowadays, the need for accuracy is no less, but the combined effect of simpler stair design, factory production using CNC (computer numerical control) processes and CAD/CAM (computer-aided design and manufacture), and advance ordering to meet contractual commitments, has resulted in the stairwell – more often than not – being formed to suit the staircase, rather than the other way round.

In either case, the following points should be given attention:

## Total rise of stairwell

*Figures 6.28(a)(b)*: To obtain truly level treads, vertical risers and newel posts, the 'total rise' of the stairwell must be divided precisely by a number of equal divisions for risers. As first illustrated in Figures 6.1(c) and (d), increases or decreases to the 'total going' – although also critical in most stair designs – do not affect the stair being level or plumb. Figures 6.28(a) and (b) illustrate the use of a timber batten known as a 'storey rod' that was traditionally used for checking the total rise of an established stairwell. Once placed in position and marked at the top, it can then be checked more carefully with a rule than a rule might convey with in-situ measuring. The rod – if straight and parallel – can also be used to plumb down the face of the trimmer- or trimming-joist for the purpose of checking the total going of the stair.

Note that the required storey-height for a pre-designed stair usually depends on the blockwork (or brickwork) being carefully controlled up to the bearing-height for the floor joists or joist-hangers

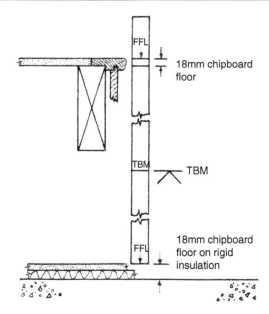

Figure 6.28 (b) When checking 'total rise', take the different, finished floor-levels into account, if – as is likely – the floors have not yet been laid.

– and relates to an upper and a lower ffl (finished floor level). To check the ffl at certain critical points like doorways and stairwells, a *temporary bench mark (TBM)* – a short horizontal line underscored with an arrow-head – is sometimes marked at doorways and stairwells, as reference points set at 1m above ffl. Carpets are discounted, but if a sand-and-cement screed is to be overlaid with a parquet floor or quarry tiles, etc, then the top surface of these is the ffl.

## Determining the bulkhead position

*Figures 6.29(a)(b)*: The bulkhead trimmer- or trimming-joist is that part of the stairwell-opening above a stair that – if wrongly positioned – can cause head injuries. AD K1 refers to it with their Headroom rule of NLT 2m above the pitch line.

Although forming the trimmed stairwell opening in a timber-joisted floor is usually done by carpenters, not joiners, such related work should be part of a joiner's knowledge – so certain points are covered here. To determine the critical position of a stairwell's bulkhead (apart from taking scaled measurements from the architect/designer's drawing), a simple calculation of the known step sizes, graphically illustrated below, can be made.

Start with the headroom requirement of 2m above the pitch line, add 40mm for tolerance and add the bulkhead's floor-depth of, say, 255mm (being 225mm joists + 18mm chipboard + 12mm ceiling). This gives a total of 2.295m to be divided by a known rise of, say, 180mm; which gives 12.75 divisions, i.e. **12¾ risers**.

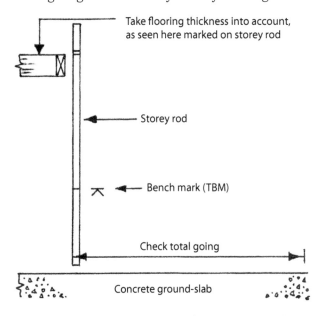

Figure 6.28 (a) Using a storey rod to check the total rise and the total going.

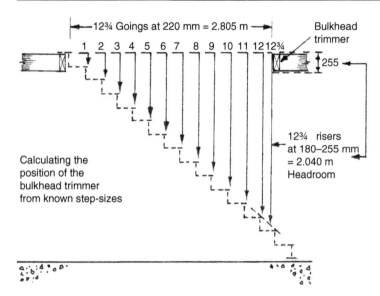

Calculating the
position of the
bulkhead trimmer
from known step-sizes

12¾ Goings at 220 mm = 2.805 m

1  2  3  4  5  6  7  8  9  10  11  12 12¾

Bulkhead
trimmer

255

12¾ risers
at 180–255 mm
= 2.040 m
Headroom

Figure 6.29 (a) A simple calculation to establish the bulkhead-position for Headroom, if forming the trimmed stairwell-opening.

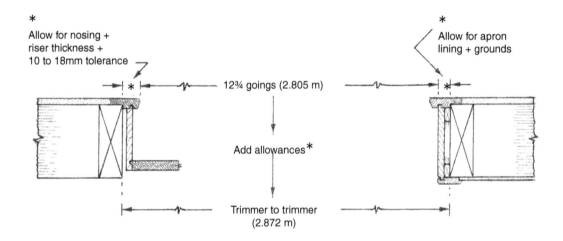

*
Allow for nosing +
riser thickness +
10 to 18mm tolerance

*
Allow for apron
lining + grounds

12¾ goings (2.805 m)

Add allowances*

Trimmer to trimmer
(2.872 m)

Figure 6.29 (b) Graphic illustration of the various allowances to be taken into account between the landing trimmer- or trimming-joists to establish the bulkhead position.

12¾ risers produces **12¾ goings** of, say, 220mm. Therefore 12.75 × 220mm = 2. 805m as the measurement of the stairwell opening from top-riser face to the bulkhead position.

## Allowances for half-turn landings

*Figures 6.30(a)(b)(c)*: The traditional half-turn stair with a half-landing in its mid-rise area was known as a 'dog-legged stair'. This was because there was no space between the two flights and the tenons of the outer string (the inclined board, housing the ends of the steps) of each flight were housed in the same newel post, giving the visual effect of a continuous dog-leg shaped outer string.

## Stairwell wall surfaces

When measuring pre-formed stairwells to confirm the height, length and width for the staircase to be made or ordered, also check that the blockwork or brickwork surfaces of the stairwell are straight, plumb, square-cornered and without any mid-area irregularities. Check also whether the bare stairwell-surfaces are to be plastered with float-render-and-set (skimmed) plaster or dry-lined with so-called 'dot-and-dab' plasterboard. This is most important because wall-strings were traditionally fixed directly to the structural wall before plastering, but such fixings nowadays make it more difficult for dry-lining operations – and there is evidence that stairwells and stair-walls are being dry-lined before the stairs are fitted and fixed.

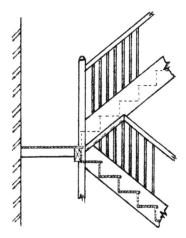 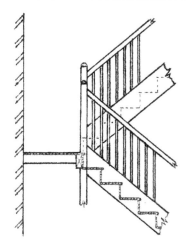

**Figure 6.30 (a)** The original features of a 'dog-leg' stair, with the lower and upper outer-strings mortised into only one newel post. As illustrated, this created a now- unacceptable discontinuous handrail to the lower flight. Hence, the introduction of double-width newel posts, or two posts side-by-side at the half-landing (as indicated in Figure 6.27 and 6.30(b)). Note that the working-out for positioning the half landing shown at (c) is not affected by double newel posts.

**Figure 6.30 (b)** Half-turn stair with two newel posts side-by-side – and continuous handrails. Note that the newels might be capped off on the underside, or continue down to the floor. Also, they might be the same height, or stepped, as shown.

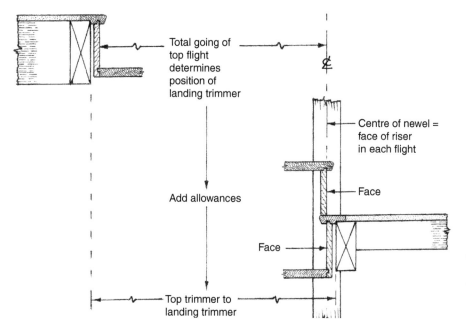

**Figure 6.30 (c)** Graphic illustration of the various allowances to be taken into account between the top landing trimmer-/trimming-joist and the mid-landing trimmer-/trimming-joist of a half-turn stair.

# DESIGNING, SETTING OUT AND MAKING STAIRS

After a stair is designed to suit a particular stairwell, either by computer-aided design or in the form of a hand-crafted, scaled plan-view drawing, the techniques and procedures of setting out and constructing are very similar and are separately covered here.

However, stairs nowadays can be made by computer-aided manufacture (CAM), or with the aid of CNC routers. My aim here is to cover the making of stairs by using hand-tools, stair jigs and portable-powered routers aided by access to standard fixed machinery such as a surface planer/thicknesser, mortising machine, band saw, etc.

## Designing a straight flight of stairs

The logical designing approach is to first divide TR (the *total rise*) by a trial-and-error intelligent number – say between 12 and 16 – until an acceptable and permissible rise has been found. For a Category 1 stair in a dwelling house, we know that the regulation rise should be *not more than 220mm* and we can be further guided by knowing that a rise of between 7 and 7½ inches (178 and 190mm) was considered to be ideal traditionally.

Next, we must consider TG (the *total going*). As this dimension might include the going of a landing (the landing's width), this is deducted from TG until the horizontal dimension between the faces of the bottom- and top-risers remains. This resultant dimension is then divided by the number of goings, which will be (remember the rule) *one less than the number of risers* – and this will determine the step-size of the *going*. Once the rise and the going are established, the degree of pitch can be checked easily, speedily, yet accurately by drawing a precise, large-scaled (1:2) right-angle to the rise and going; then by forming a hypotenuse to the extremities of these lines, thus enabling the *pitch* to be measured with a protractor (preferably the adjustable, set square type). With a 1:2 scale, this can be done on A4 size paper.

Of course, if you have a preference for trigonometry and a scientific calculator, the pitch angle equals the *opposite* (say, a rise of 192mm) divided by the *adjacent* (say, a going of 237mm), all multiplied by the shift tangent. Such calculating results in a pitch angle of 39.01°. My preferred method with a scaled triangle produces a pitch angle of 39° – close enough to the trigonometry and acceptably below the regulatory maximum of 42°.

## Graphic illustration of basic design

*Figures 6.31(a)(b)*: These cross-lined elevational grids serve to illustrate graphically the simple mathematical relationship that exists between the division of *total rise (TR) and total going (TG)*, of (a) a straight-flight stair and (b) a half-turn stair. Note that as illustrated in the two-flight, half-turn stair, instead of *one less going than the number of risers*, there must be *one less going than half the number of risers*. Note also that the stairs illustrated here are visually more representative of concrete stairs than wooden stairs – this is because the simpler appearance of concrete stairs seemed to make my graphs more understandable – but the theory of TR- and TG-division is the same.

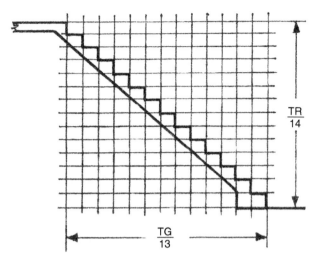

Figure 6.31 (a) 14 risers produce 13 goings (because the top landing is excluded).

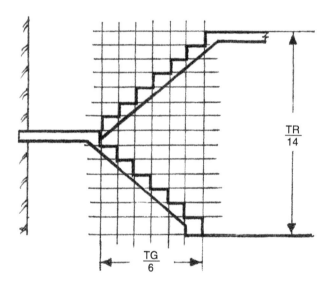

Figure 6.31 (b) This design of a half-turn stair with 14 risers must have one less going than *half* the number of risers.

## Setting out from drawings or site measurements

For the purpose-made staircase, a joiner or woodcutting machinist takes available information from the drawings and contract specification. But – if possible – a site visit is advisable. When the timber is available, he proceeds to calculate the precise step size – the rise and going. As an example, assume that 13 treads (goings), excluding (as previously explained) the top landing-step, and a total rise of 2.688m were indicated on the drawing, the sum would simply be 2688 ÷ 14 (13 treads plus 1 for the top landing-step) = **192mm rise.**

## Practical dividing method

*Figure 6.32*: Traditionally, this sum was arrived at more tediously. The total rise was marked on a storey rod (a timber batten) and the rise was stepped out with a large pair of carpenter's dividers (Figure 6.32) by a trial-and-error method of repeated mini-adjustments, until the exact division of risers was determined. This practical method of division was quite accurate, but relatively time-consuming. However, it must be mentioned that the division of a total rise of 2.688m (used above) was in fact approx 8ft 9¾ inches in pre 1973 imperial measurements. Even though the feet could be easily converted into inches, most divisions required the inches to be converted into fractions. For example, the rise (used above) of 192mm would be approx *seven inches and thirty-five sixty-fourths of an inch*. Knowing how people generally feel about fractions, might explain why a practical method of division evolved in yester-year before metrication and calculators!

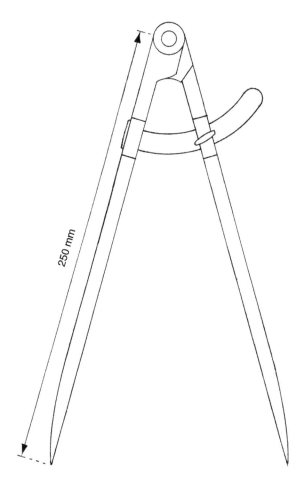

**Figure 6.32** A pair of wrought-iron dividers with wing-nut adjustment, used traditionally by carpenters and joiners for setting out. Note that the usually dull points of the dividers were easily filed with a mill file, or saw file until sharp points were produced.

## Main stair components required

### Treads

Nowadays, 18mm MDF (medium-density fibreboard) is often used for treads, but it is prone to short-grain delamination when grooved near its nosing edges to receive the tops of risers and tends to split on its back edges when receiving screws from the riser fixings. On better-class work, treads are of timber (mostly softwood, but hardwood stairs are also popular), usually varying in finished thickness between 18 to 28mm.

### Risers

Traditionally, these were also of timber boards (and still may be if the stair is to be made in hardwood), usually of 16 or 18mm finished thickness. In the 1970s, 9mm- or 12mm-thick plywood was used quite effectively – but nowadays, this appears to have been mostly displaced by the use of 9mm or 12mm MDF (which is cheaper) – although there is evidence that plywood is also still in use.

### Strings

These are the load-bearing, pitched on-edge boards that house and carry the ends of the steps on each side of a stair. For obvious reasons, one is named a *wall string* (or *inner string*) and the other is named an *outer string*. The width of these boards can vary to suit the size of the steps and the size of the 'margin' below the string's top edge (explained further on). However, the common minimum width of string is usually ex 225mm (220mm finish) – although ex 250mm (245mm finish) gives a more generous margin. The common minimum thickness is usually ex 32mm (28mm finish) – and more substantial stairs (usually in Category 2 or 3; or Category 1 Stairs with *cut and bracketed strings*), can require strings of ex 38mm or more.

### Joints between treads and risers

*Figures 6.33(a)(b)(c)(d)*: If a *cutting list* of all the required components of the staircase is to be produced (usually required if the work is split between joiners and machinists), giving details of length, width, thickness, etc, it will be necessary at this stage to establish the method of edge-jointing to be used between the treads and risers. Four methods are shown below and although method (a), with a plywood riser is common practice and has a good rating in first-class softwood stairs, theoretically, there is no allowance for tread-shrinkage. When using softwood (or hardwood) boards for risers, methods (b) and (c) are regarded as the best methods to combat the possible effects of shrinkage.

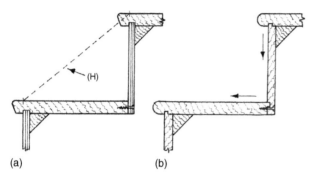

(a)                          (b)

**Figure 6.33 (a)** In theory, because the tread is only butt-jointed to the plywood riser, if it shrinks across the grain, there is a likelihood of a gap appearing against the riser face. Note that illustration (a) shows the hypotenuse of the step-triangle at (H), the length of which is used to multiply the number of steps to determine the initial length of the strings. There would of course be additional allowances for either newel-post tenons, or – in the case of wall strings – projections to intersect with the upper and lower skirting boards. Also note that the hypotenuse at (H) becomes an important 'margin line' when setting out the steps (to be covered in following paragraphs); and **(b)** this method of jointing allows each board to shrink slightly (in the direction of the arrows) without splitting or showing unsightly gaps. Such gaps might not concern the majority of people who carpet their stairs, but there are others who prefer wooden stairs – in softwood or hardwood – to be fully exposed.

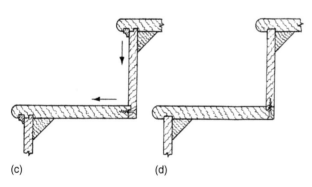

(c)                          (d)

**Figure 6.33 (c)** This method of jointing is similar to (b), regarding shrinkage-design, but the nosing appearance is made more ornate by the insertion of a so-called 'scotia mould' (traditional terminology for a 'cavetto mould'); and **(d)** this method of jointing allows the riser to shrink, but not the tread board.

## Glue blocks

As shown in Figures 6.33(a) to (d), triangular blocks (referred to as 'glue blocks') are an important part of the structure of all wooden or composite-boarded steps and they are produced by ripping through a length of 50 × 50mm par timber diagonally (with the aid of a saw-bench jig), prior to crosscutting to

length. They are recommended by BSI (the British Standards Institute) to have a sectional size of not less than 38mm × 38mm and a minimum length of 75mm. For stairs up to 900mm wide, at least two blocks per step should be glued and rub-jointed to the internal angle between each tread and riser – at the separate stage of step-construction (described later).

## Timber quality

If the quality of the timber is to meet the British Standard, BS 585 refers to BS 1186: Part 1: class 3, which list requirements of grading. This will mean careful selection of the staircase timber, as class 3 does not refer to commercial grading descriptions.

## Making step templates

*Figures 6.34(a)(b)(c)(d)(e)*: Although a traditional-type steel roofing square with adjustable stair-gauge fittings – Figure 6.34(a) – was effectively used to set out the steps on stair strings (especially by site-carpenters setting out formwork strings for in-situ concrete stairs), not all joiners had these tools and – prior to modern *plastic-laminate stair-template jigs*, now widely used with heavy-duty portable powered routers – wooden-, then plywood-, then hardboard-templates were made for setting out the exact step positions on the stair-string boards. As the updated template is still used, the knowledge-related progression of these is shown below, from 6.34(b) to (d).

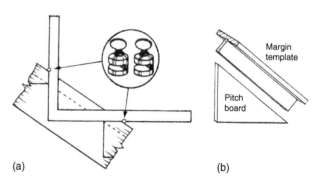

(a)                          (b)

**Figure 6.34 (a)** A steel roofing square (now called a metric rafter square, sized 610 × 450mm) in position – with a set of separate stair-gauge fittings (still available) attached to the 'tongue' and the 'blade' seated on the face of a stair string being marked out. Note the dotted line representing the margin needed to accommodate the tread-nosing projections; and **(b)** a triangular pitch-board with a separate margin-template. As seen in Figures (c) and (d), the margin template eventually became a part of the pitch-board.

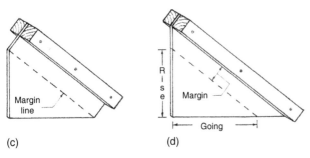

(c)                              (d)

**Figure 6.34 (c)** The combined pitch-board and margin template; and **(d)** the updated version of the pitch-board, made of thin plywood, hardboard or 4mm MDF. Note that the extended rise- and going-edges allow the pencil to run on and – once the pitch-board is moved to the next consecutive step-position – the crossed lines create a better intersection at the critical extremes of the hypotenuse on the margin line.

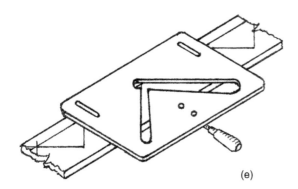

(e)

**Figure 6.34 (e)** An *isometric impression* of a Trend 'A' plastic-laminate staircase jig clamped in position on a marked stair-string, ready for the step-housings to be formed with a heavy-duty portable powered-router.

## Preparation of material for steps

After selection of the timber and the necessary machining – including surface- and thickness-planing – the treads and riser boards should be cut accurately to length. This can be done by hand or on a dimension saw-bench, if available. The edge treatment of tongues and/or grooves and the nosing-shape is next. Again, these can of course be done by hand with a variety of traditional planes, otherwise a vertical spindle moulder machine or a heavy-duty portable powered-router is needed.

## Preparation of Strings

*Figure 6.35*: Next, the inner and outer strings need to be prepared for housing or trenching to receive the steps. Traditionally, in the absence of powered routers,

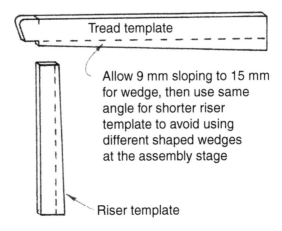

Tread template

Allow 9 mm sloping to 15 mm for wedge, then use same angle for shorter riser template to avoid using different shaped wedges at the assembly stage

Riser template

**Figure 6.35** Over-long Tread- and Riser-templates.

the housings would be cut by hand methods (referred to as 'trenching') and two more templates would be required. The first, illustrated in Figure 6.35, is a tread template, which includes the nosing shape, the exact projection past the face of the riser and the wedge shape. The second template in Figure 6.35 is for the riser, which also includes the wedge shape. The templates' lengths purposely exceeded the actual width of the treads and risers slightly, beyond the bottom edge of the strings to enable the marking out of the housings to run-off the edge.

## Template details

Even though only the pitch-board template described at 6.34(d) above is likely to be used nowadays for marking out the stair strings prior to attaching a staircase jig, it should either be made with good-quality 6mm plywood, a dense grade of 3mm hardboard, or 4mm MDF. It must be carefully prepared to shape and size, as accuracy is vital in setting out – especially with a pitch board; even 0.5mm more or less in riser-height will have a cumulative effect of a 7mm error on the total rise of a stair with 14 risers.

The top of the pitch-board above the margin allowance must be fitted with a double, wooden fence – as at 34(d) – to rest against the top edge of the strings. The fence material, of not-less-than 20 × 20mm prepared sectional size, can be glued and pinned, or screwed – but must be accurately positioned in relation to each other.

## Margins

*Figure 6.34(d)*: The margin indicated in Figures 6.34(a) to (d) is a variable measurement, usually set at not-less-than 50mm, but it is dependent on the width of timber being used for the strings and the edge-treatment required, if any, on the top edge of the inner wall-string – i.e. if to be moulded to meet a moulded skirting-shape. However, the margin should comfortably accommodate the nosing projections without bringing them nearer than 25mm to the string's edge.

## Setting out the stair strings

*Figures 6.36(a)(b)(c)(d)* and *Figure 6.37*: When setting out the stair strings for the all-important step-housings, first the margin line is pencil-gauged along the face-side of the string. Then the pitch-board is positioned at the base of the wall string to mark the floor-line and the first riser of the bottom step. A pair of dividers (as shown at Figure 6.32), or a beam-compass, is opened out along the margin line and is carefully set between the intersections at finished-floor-level (FFL) and at the top of the riser. This is the true hypotenuse of the step (if we ignore the nosing projection), as clearly indicated at (H) in Figure 6.33(a). The dividers are then carefully stepped out along the line by the number of steps required. The pitch-board is used again and the face lines of the treads and risers are pencil-marked – care being taken to relate each position of the pitch-board to the equidistant divider-marks along the margin line.

Figures 6.36(a) and (b) illustrate the broken-length parts of a wall string; with (a) the pitch-board correctly positioned above the marked-out steps (the only marking out required if using a powered router); and

(b) tread- and riser-templates in position at the top end of the string to illustrate additional marking out required if the housings were being trenched out by traditional hand operations. Figure (c) shows the finished appearance of the lower part of the string after trenching or routering; and (d) shows the upper part of the string with the top step in position.

Figure 6.37 shows a reliable method of transferring the stepped-out divider marks from the wall string to the outer string. This saves time and also reduces the risk of cumulative errors that may arise from setting out the strings separately.

## Marking out the string tenons

*Figures 6.38(a)(b)(c)(d)(e)(f)*: The outer string seen below in Figure 6.37 shows the position of the oblique stub-tenons for jointing to the blind mortises in the newel posts. The tenons on the right relate to the top riser and the newel post positioned at the landing level; the tenons on the left – which would normally relate to the bottom riser if there was no protruding step beyond the bottom newel post – relate, as shown, to the second riser up, such being the case if (as was quite common) the bottom step protruded past the newel post.

Single, protruding bottom steps, with a traditional bull-nosed end, a modern square-end, or a 45° splayed end (as in Figures 6.24(b), 6.27 and 6.43), regardless

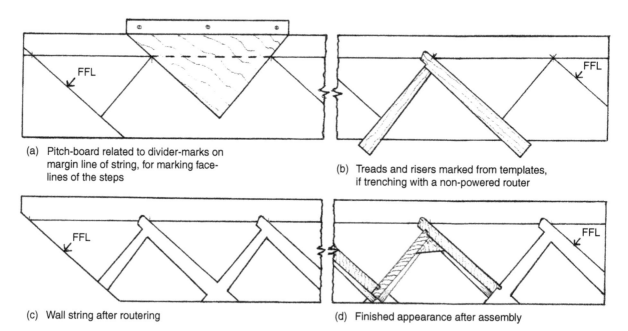

(a) Pitch-board related to divider-marks on margin line of string, for marking face-lines of the steps

(b) Treads and risers marked from templates, if trenching with a non-powered router

(c) Wall string after routering

(d) Finished appearance after assembly

Figure 6.36 (a) Pitch-board in position at lower end of wall-string, one up from FFL – finished floor level; and (b) tread and riser templates in position at upper end of wall string – one down from FFL; note the allowance left above FFL, to be trimmed on site to suit the skirting height; (c) the appearance of the routered housings and the waste removed below FFL (though this and the skirting abutment are done on site); (d) an illustration of the last fitted and wedged step at the top of the wall string – note that the very top riser and the landing nosing-piece are also fitted and fixed on site.

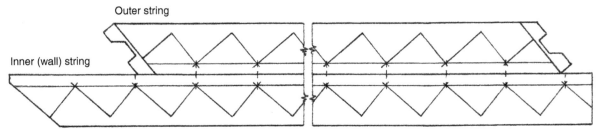

Margin-line marks squared across from the wall string to the outer string, prior to pitch-board marking

**Figure 6.37** This illustration of a pair of stair-strings positioned edge-to-edge highlights the method of transferring the hypotenuse divider-marks from the wall-string to the outer-string – and also shows the position of the outer-string's tenons, which have evolved from the pitch-board's initial face marks – explained in the setting out details in Figures 6.38(a) to (e). It can also be seen here (on the left) that the inner wall-string runs past the bottom newel-post tenons, this being for the inclusion of a bottom bullnose-ended step that protrudes past the newel post.

of any perceived ugliness or aesthetic appeal, add greater strength and rigidity to the post. This is because the post is supported higher up by the string's tenons and by the screwed end of the second riser. In marking out the oblique stub-tenons, the position of the shoulder line is determined by the key rule that *the face of the riser* (marked on the strings from the pitch-board) *equals the centre of the newel.* Therefore, as indicated in Figure 6.38(a), the shoulder line is marked back (into the string) at half the newel's thickness from the riser-face, then the tenon-projection is marked forward at two-thirds the newel-thickness from the shoulder line. After the bare-faced tenons have been shouldered to thickness (usually 16mm for a 28mm finished string), the double stub-tenons are set out as shown below in Figures 6.38(a) to (e).

## Removing tenon-cheeks from strings

*Figure 6.38(b):* The deep, oblique stub-tenon area – as seen free of marking out at 6.38(b) – is reduced on the step-side by 12mm to leave a bare-faced tenon area of 16mm. This reduction can be done on a tenoning machine, by using either the bottom- or top-tenoning head – otherwise such a deep, oblique tenon is awkward to reduce and various techniques are used to achieve this, including: 1) Reduce the area to a 12mm depth by a series of portable powered-router cuts; 2) Cut the shoulder-line with a tenon saw, remove about an 8mm depth of waste wood by chisel-chopping across the short end-grain (providing the grain is straight), then (having now produced a shoulder to run against) remove the remaining waste down to the gauge lines with hand tools such as a Stanley No.78 rebate plane and/or a No.073 shoulder plane, finishing in mid-area with a Stanley 04 or 04½ smoothing plane – or a 5½ jack plane.

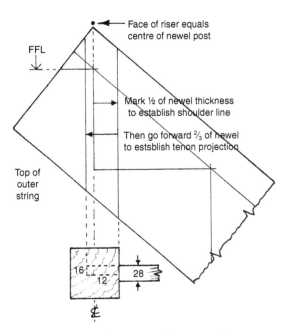

**Figures 6.38 (a) to (f)** The *part elevations* of the outer string at the top-end, above a plan view of the newel post and bare-faced, stub-tenoned string in-line below, shows the step-by-step setting out of the oblique, double tenons required. The same procedure for this setting out can also be applied to the tenons at the bottom of this string. Note, however, that when there is no protruding bottom step, the outer string touches the floor and the tenon needs to be divided by 4 instead of 3, to provide a base haunch. Note also that, as shown at (f), these wide double-tenons theoretically should have a 5mm reduction at their extremities (back to the shoulder line) to guard against the mortise holes appearing via string- and tenon-width shrinkage. However, to my knowledge, this is rarely done; partly because the top edge of the outer string is covered with a grooved string-capping (to receive the balusters) and partly because the acute angle formed on the underside abutment to the newel makes it virtually impossible to detect any shrinkage gaps.

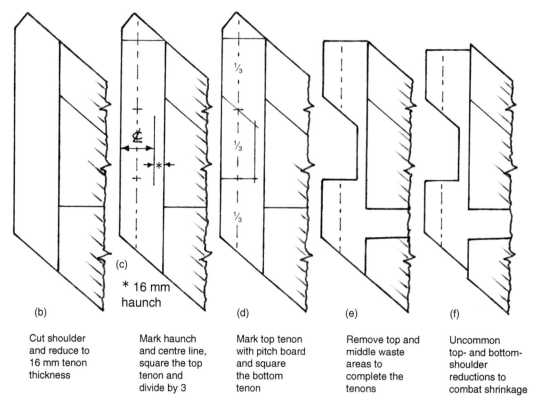

(c)

* 16 mm
haunch

(b)

Cut shoulder
and reduce to
16 mm tenon
thickness

Mark haunch
and centre line,
square the top
tenon and
divide by 3

(d)

Mark top tenon
with pitch board
and square
the bottom
tenon

(e)

Remove top and
middle waste
areas to
complete the
tenons

(f)

Uncommon
top- and bottom-
shoulder
reductions to
combat shrinkage

Figures 6.38 (a) to (f) (continued)

## Housing the strings by traditional hand-techniques

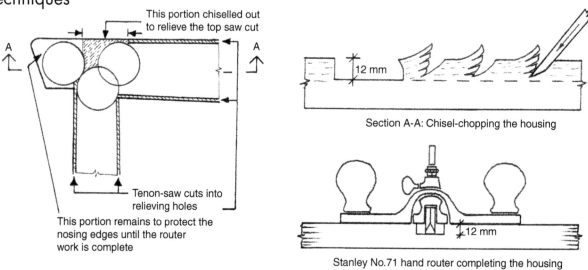

This portion chiselled out
to relieve the top saw cut

Tenon-saw cuts into
relieving holes

This portion remains to protect the
nosing edges until the router
work is complete

Section A-A: Chisel-chopping the housing

Stanley No.71 hand router completing the housing

Figure 6.39 The above procedure was necessary when trenching out the step-housings by hand-skills techniques. Relieving holes of 12mm (or less) depth were made with a Forstner-type bit to receive the toe-end of a tenon saw and the sides of the housings were cut to an approx angle of 1 in 12 (86°); this was to achieve a dovetail effect when the steps were wedged up. Then the housings were chopped up quite deeply with a firmer chisel and mallet – and finished off with the hand router.

*Figure 6.39*: It may be of interest to enthusiasts of traditional hand-skills – as opposed to mechanized hand-skills – to glimpse the housing-technique used in small joinery shops about five decades ago, before portable powered routers became established (even though the first portable router (the 'Kelley Electric Router') was patented and marketed in Buffalo, USA in the early 1900s). As illustrated, the string-housings

– after being marked out with tread- and riser-
templates – were counterbored with three holes of,
say 25mm Ø (diameter) near the nosings. This was
done to facilitate the tenon-saw cuts after paring out
the shaded portion shown above. The hand-router work
was greatly relieved by chopping quite deeply across
the grain with a firmer chisel and mallet – illustrated in
section A-A.

## Housing the strings by machines

Nowadays, the stair-string housings are done by a
variety of machines – apart from heavy-duty portable powered routers. Such machines include CAD –
CAM (computer-aided design and manufacture)
production and the use of CNC (computer numeri-
cally controlled) routers.

## Using handmade jigs

*Figure 6.40*: The heavy-duty portable powered plunge-
router requires controlled guidance, either in the
form of patented stair-template jigs, as illustrated in
Figure 6.34(e), or in the form of hand-made plywood
jigs, such as that illustrated in position in Figure 6.40.
The jig is either clamped or pinned to the string at
each consecutive move – and requires its wooden side-
fences to be repositioned (turned over) for the jig to be
operational on the opposite (left- or right-hand) string.

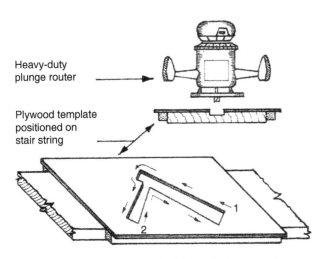

Heavy-duty
plunge router

Plywood template
positioned on
stair string

Figure 6.40 A purpose-made left-handed plywood jig in
position on part of a stair string, with a portable plunge-
router shown above it.

## Nosing-shape adjustment

*Figure 6.41*: If the required nosing-shape for the treads
is different to the semi-circular end of the routered
housings, as indicated in Figure 6.41, a small portion

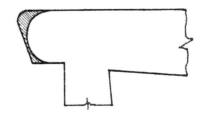

Figure 6.41 The shaded area, which is marked from the
step's profile after the routered housings are made, indicates the portion of string to be shaped by chisel-paring.

of the machined housing will have to be finished by
hand-chiselling techniques – i.e. vertical paring with
bevel-edged chisels and in-cannel scribing gouges.

## String tenons and dowels

*Figure 6.42*: The stub tenons formed on strings for
insertion into the blind mortises in the newel posts,
seem to vary in depth from a half to two-thirds the
post's thickness. I believe the latter is better. This is
because these tenons (used also for handrails) are usually
dowelled by a technique known as 'draw-boring' and the
grain of the shorter stub tenon (as illustrated in Figure
6.42) has a tendency to shear if the tenon is too short.

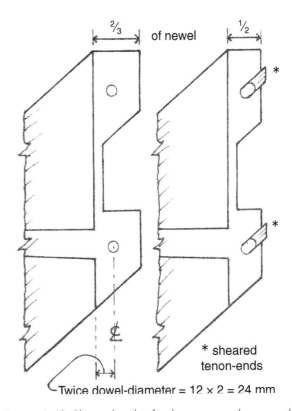

$\frac{2}{3}$ of newel     $\frac{1}{2}$

* sheared
tenon-ends

Twice dowel-diameter = 12 × 2 = 24 mm

Figure 6.42 Sheared ends of stub-tenons can be caused
by draw-bored dowels if the tenons are only half the thickness of the newel post.

There are two reasons for dowelling tenons in stair work. 1) The tenons are not wedged and may become loose if shrinkage occurs and 2) by draw-bored dowelling (offsetting the drilled dowel-hole by about 2mm in the tenon *only*, towards the shoulder of the newel post), the shoulders of the joints can be pulled up when the blunt, pencil-like sharpened dowel is driven in. It also has to be understood that a rhomboidal-shaped frame of two parallel newel posts and a handrail parallel to the outer string, cannot be cramped together (especially on site) in the normal way that a square-shaped frame is cramped.

The traditional rule for draw-boring is that the centre of the positioned dowel should be twice its diameter from the shoulder of the joint – as illustrated above.

## Marking out handrails and newel posts

*Figure 6.43*: Before assembling the stair flight, the shoulders of the handrail tenons are best marked from the shoulders of the outer-string tenons, by laying and cramping the handrail along the string's top edge. Also before stair-assembly, the step-housings and

blind-mortises in the newel posts should be marked. This can be triggered off by laying each newel tightly against the shoulders of the string-tenons and haunching, then by pencilling their outline onto the newels. These outlines can then be squared across the inner, adjacent face of the newel to give the mortise and haunch positions. This technique can also be applied to the handrail-tenons (See Figures 6.43c and d).

After the newels have been mortised, all tenons should be tried for fit. At this stage, the draw-bored holes in the newels, strings and handrails can be done – rather than leaving this task to the site carpenter – and the drawn-up quality of the shoulder-fits can be checked by driving in 12mm Ø tapered metal draw-bore pins, hooked at their ends for easy removal (as shown in Chapter 3).

Figure 6.43 A, B, C and D illustrate the marking out on the four unfolded faces of a bottom newel post to receive a protruding step housed into its front face. As shown below FFL, an allowance is normally given at the bottom of newels to cater for the various methods of site-fixing. The tops of the newels may or may not be finished, depending on whether they are finished in themselves, or are to have separate caps fitted on site.

Riser- and tread-housings into strings are commonly 12mm deep, but because the risers into newel-housings are screwed through the risers' back-side – and not wedged – the housings in the newels should be 18mm deep.

## Stair assembly

*Figures 6.44(a)(b)(c)*: Assembling a stair flight has always presented a problem in the gluing-up

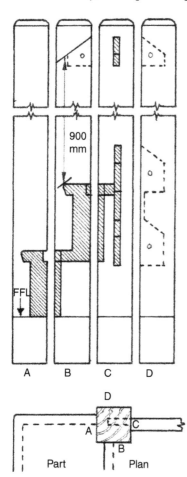

Figure 6.43 A, B, C, D: *Elevation* and *part plan* of unfolded newel-post faces.

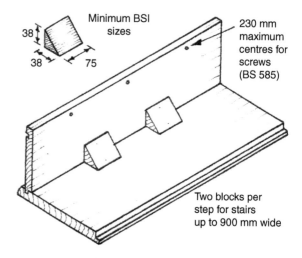

Figure 6.44 (a) The tongued riser is glued to the tread groove, then at least two blocks are glued and rubbed into position – and pinned, if necessary, to achieve a true right-angle after checking the inner L-shape with a try square.

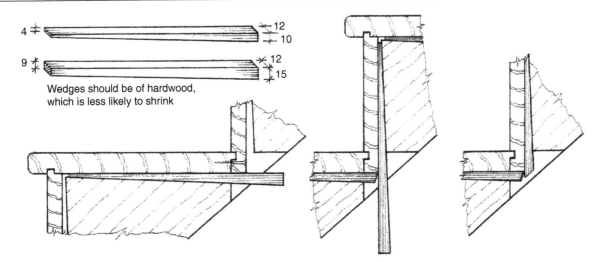

Wedges should be of hardwood,
which is less likely to shrink

**Figure 6.44 (b)** After knocking up the treads to fit the nosing-housings tightly (with a claw hammer and 'hammering block'), the tread-wedges are the first to be glued and driven-in. In practice, this is best done more thoroughly and speedily by generously brushing glue into the housing (and underside of tread), inserting the wedge, pressing it in against the string with the finger-tips and then hammering it in. After the tread-wedges are trimmed to allow entry for the riser wedges, glue is applied again and they are driven in as before. The trimming of the wedges, as described above, completes the gluing-up and stair-assembly.

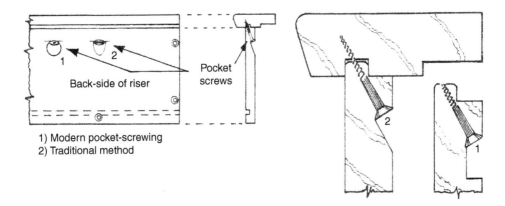

Back-side of riser

Pocket screws

1) Modern pocket-screwing
2) Traditional method

**Figure 6.44 (c)** Note that the top riser of a flight of stairs and the 'top tread' (the latter known in the trade as a 'landing-nosing' or 'nosing-piece') – both illustrated above – are not fixed in the joinery works. This is partly because the rebate of the nosing-piece often requires 'easing' to suit the floor thickness, but mostly because one end of the riser needs to be screwed to the newel-post housing – and the newel(s) are not fixed at the joinery stage. The reasons for this are for easier transportation, manoeuvrability through doorways and other practical issues involved in the fitting and fixing. In total, stairs usually arrive on site separated from the newel posts, the balustrade, the bottom step(s) (if one or two protrude beyond the newel), the top riser, the landing nosings and the apron linings.

stage – although this is less so with modern adhesives. However, contrary to the recommendations of BS 585:1972, in my experience the practice has always been to dry-house the steps into the strings and only glue the wedges.

To improve on this traditional practice, the following method has been tried and proved to be successful – but is more time-consuming.

After cleaning up (sanding) all visible surfaces, each tread and riser are glued together to form separate

steps; two or three angled glue-blocks are glued and rub-jointed into position – as illustrated in Figure 6.44(a) – and, if necessary, these may be fixed with panel pins whilst testing the upstanding riser board of each step with a try square. The steps are set aside in one or two stacks of riser-up, riser-down formation.

After allowing for the glue to set, preferably until the next day, the steps are positioned in the dry housings of the wall-string laid face up on a stair-assembly bench and the opposite (outer) string is placed on top.

Directly above this, a cramping arrangement – either of telescopic trench-props, improvised timber struts with folding wedges, or (in large workshops) permanent screwing jacks connected to an overhead beam – must be in place to exert a downwards thrust on the encased steps.

Once this is set up, only a minimum of pressure is first applied, then the steps are eased open at the unconnected joints between the back of the treads and the bottom of the risers, glue is worked in, the joints are brought back together and screws are power-driven into previously-drilled shank holes.

When the complete flight of steps is screwed together, the pressure is released from the cramping arrangement and the top string is removed and replaced again after applying glue to the housings. The reassembled flight is then carefully, but quickly turned over, the dry-housed string on top is removed, glued, replaced, and pressure (more than at the start) is reapplied from above.

The treads must be quickly knocked up tightly into the housed nosings (with an offcut of wood – a 'hammering block' – taking the hammer blows); then each of the tread wedges is glued and driven in, followed by the riser wedges (after cutting off the tread-wedge projections with a chisel and mallet). To complete the operation, the riser wedges are trimmed off by the same method – by inclining a wide chisel, with its ground- and sharpened-edge against the underside step-arris and striking one or two blows.

Note that as the sides of the string-housings are slightly bevelled (this being automatically achieved by a shaped TCT router cutter), when the wedges are driven in, a dovetail effect is achieved. It should also be noted that some joiners prefer the wedges to be more robust; say, instead of the tread-wedges being 10mm diminishing to 4mm, they may be 15mm to 9mm. But, remember – to avoid a mix-up of wedges – the riser-wedges (and riser-template, if used) must be to the same angle.

It should also be realized that the top riser and the landing-nosing – because they are up against (or very close to) the landing trimmer-joist – cannot receive glued angle-blocks. For this reason, it is most important that the site-fixer glues and screws this joint with at least three so-called 'pocket screws'. In Figure 6.44(c), the recessed niche for the pocket screw is shown traditionally gouged out by an in-cannel scribing gouge, but, if preferred, suitable recesses can also be formed by drilling shallow holes with a Forstner-type bit of, say, 16mm Ø, prior to drilling the angled shank holes.

# Tapered steps and their shaped strings

*Figure 6.45*: Arrangements of tapered steps and their inner wall-strings are best set out on a 'rod'. As described in Chapter 2, rods are full-size drawings of elevational, plan or sectional views that develop and/or display the details required. The plan- and elevational-view of one of the developed wall strings seen below was set out on sheets of hardboard that had been painted with white emulsion.

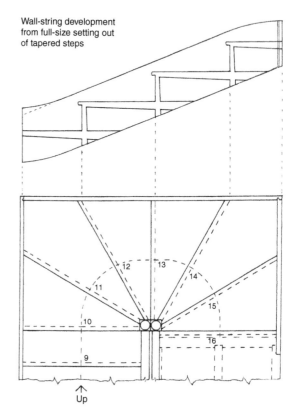

Wall-string development from full-size setting out of tapered steps

Up

Figure 6.45 This half-turn stair of tapered steps (taken from Figure 6.27) was set out to full size on a painted hardboard rod. First, the two newels were drawn in a central position, so that nosing-lengths 10, 13 and 16 were equidistant to each other. Then the nosings were set at 30°, 60° and 90° from the left, then the right, radiating from a central point at the back of the newels. The quick, easy and most accurate way to do this is to describe a semi-circle on the rod (with a beam compass in the form of trammel heads) to the flight's width, radiating from the newels' central back-point. This semi-circle will look similar to the 'deemed width' line in Figures 6.6 and 6.23(b) – although those are only quadrant shaped. Then set the compass to an intelligent guess and make trial-and-error adjustments until three exact steps fit around half the semi-circle, between nosings 10, 11, 12 and 13 – or between nosings 13, 14, 15 and 16. Mark them and draw the nosing-lines back to the newel's radial point. Once the riser faces are set back and drawn on the rod's plan view, the points that touch the wall string are extended out (as shown by broken lines) and the riser heights added to plot the wide ends of the steps – which produce the string shape.

## Easing of Strings

*Figures 6.46(a)(b)*: As can be seen in Figures 6.45 and 6.46, when an arrangement of tapered treads meets the continuous wall string or strings that house the parallel treads of an integrated flight of stairs, the line of nosings spreads out and either breaks through the top or the bottom of the string. This is more noticeable in Figure 6.46, where the primary positions of the strings' edges are shown as unbroken- and broken-lines. Of course, these wider housings are not allowed to actually 'break through' and, where necessary, the strings are built up in width (edge-jointed) and these built-up edges are shaped to form so-called 'easings'.

An easing can be a concave or convex shape and is first plotted when the strings of tapered treads are developed from the plan view drawn on the white-painted rods. By keeping the established top-margin in mind, a flexible piece of material (a narrow strip of hardboard, plywood or plastic trim) is used as a flexi-curve drawing-aid to mark the desired easing-shape. This is best done with one person flexing the drawing-aid to pre-plotted margin-marks, whilst another person marks the easing line.

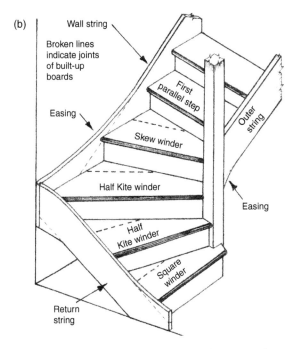

**Figure 6.46(b)** Note that the under-edge piece, beneath the left-hand, vertical tongue-and-grooved corner joint, is not shaped (because it is concealed), but concave and convex easings have been applied elsewhere.

## Tapered steps with a wreathed handrail

*Figure 6.47*: This part-elevation and plan view of a quarter-turn tapered stair, rising up from a living room and serving a galleried landing and a doorway to the upper, split-level part of the house, was an actual working drawing of a conversion job I did on a house I once occupied in 1994. Originally, the living room had a 6-tread straight stair-flight and a small landing that served only the doorway. Because of a desire to create a feature staircase, the challenge was to design a stair that met the regulations, had 6 treads and 7 risers (the same as the original, because the storey-height cannot be changed and dictates the number of risers) and last, but not least, was attractive.

After several preliminary sketches, the final layout evolved and was then drawn to a 1:25 scale (because this suited A4 sized paper) and – eventually – made by hand, with the aid of portable-powered and fixed machines – including a lathe for turning the newel posts. The balusters (or spindles, if you prefer) were bought from a parts-manufacturer and the main features of their shape were used in the newel-design. One of the unique features of the staircase was that the plywood risers and apron linings to the galleried landing were pierced with a repetitive tracery pattern, as illustrated. This pattern also pierced the bottom tapered step and its bullnose-end. Note that the short, outer string is wreathed and it has a wreathed handrail above it.

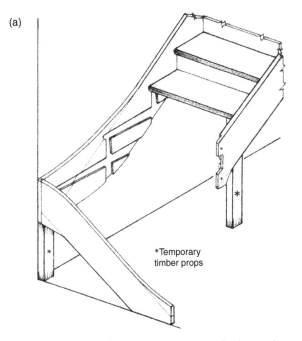

**Figure 6.46(a)** In this stair arrangement, built-up edges were required on top of the continuous wall string, on the under-edge of the outer string (close to the newel post) and on the top and under-edge of the short, return wall-string.

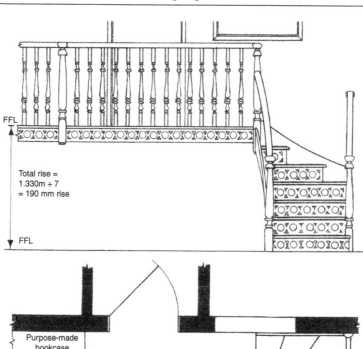

FFL

Total rise =
1.330m ÷ 7
= 190 mm rise

FFL

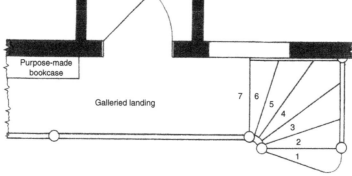

Purpose-made
bookcase

Galleried landing

7  6
      5
        4
          3
            2
              1

**Figure 6.47** (a) *Part elevation* and *part plan* of a tapered stair with a wreathed handrail above a wreathed outer string; the tracery-patterned pierced risers and apron linings are of plywood – as is the bullnose-ended tapered step, with a hand-made bent-plywood, saw-kerfed riser.

Figure 6.47  (b) The galleried landing.

Figure 6.47  (c) Close-up of the fretwork risers.

**Figure 6.47 (d)** Close-up of wreathed handrail and top edge of wreathed outer-string. (These photographs, taken in April 2013, are by kind permission of Russell and Lesley Griffiths, the present owners of the property containing the staircase.)

**Figure 6.48 (a)** Photo taken in 1995 with my grandson Luke on the bottom step – soon after the staircase was completed.

**Figure 6.48 (b)** Photo taken in 1995 with the designer/stair-builder and author in mid-flight.

## Setting out the above 6-tapered-tread stair

*Figures 6.49(a)(b)(c)*: As established previously in the detail related to Figure 6.45, one of the first jobs was to set out the full-size plan of the tapered steps (to enable the initial, oversized treads to be laid in their precise position and marked out for length and tapered shape) and the elevational-views of the strings (to enable the 'easing' curves to be developed by extending the tapered-tread extremities up from the plan-view). If possible, this should be drawn on one rod, say a sheet of white-emulsioned hardboard, but – for convenience – I did it on three half-sheets of painted hardboard, as indicated in Figure 6.49 below. Once the plan-view rod had been set out, the other two rods were joined to it to form an L shape. To avoid accidental separation, they were held in position with self-adhesive duct tape. Note that the partly-drawn newel posts were not on the original rods, but an allowance was shown past the shoulder lines for string-tenons into the posts.

First – on the plan-view rod – a large right angle representing the risers of steps No.2 and 7 (Figure 6.53(a) and shown above unnumbered) was drawn on the L/H side – its corner being the centre for all the radial riser-faces and for the quadrant-shaped deemed width and 2R+G pitch line (also shown here with broken lines). Next, the path of the deemed-width quadrant (excluding the bottom, protruding step) was divided into five by trial-and-error stepping-out with a beam compass/trammel-head compass and the intersecting arcs were lined up to the centre point and pencil-lined to represent the inner-four radiating riser-faces. The same compass setting was then marked beyond the right-hand newel position and pencil-lined back to the radial centre to form the bottom, protruding step. Next, parallel lines were drawn, projecting by 22mm from each riser-face to represent the edge of each step's nosing. The shape of the bullnose-end was then established with a compass (as in Figure 6.51(c)) to allow a short, non-curved entry into the newel post. Finally, by trial-and-error compass work – to

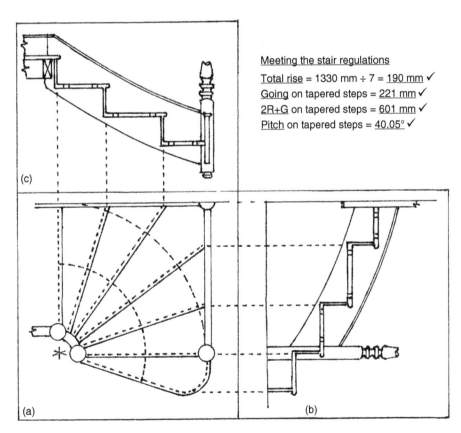

Meeting the stair regulations

Total rise = 1330 mm ÷ 7 = 190 mm ✓
Going on tapered steps = 221 mm ✓
2R+G on tapered steps = 601 mm ✓
Pitch on tapered steps = 40.05° ✓

**Figure 6.49** **(a)** The *plan view* rod; **(b)** the *elevational view* rod of the outer return-string; and **(c)** the *elevational view* rod of the inner wall-string.

enable the narrow tread widths to be checked for the required regulatory minimum of 50mm – the quadrant-shaped outer wreathed-string was drawn from the riser's centre point.

To develop the shape of the strings, the points of the risers and nosings that met the string-faces in the plan-view rod were extended out (as indicated by broken lines) onto the elevational-view rods and the riser heights were added to plot the wide ends of the steps. Once these wide-ended steps were drawn, top and bottom margin-marks were pencilled in tangentially to the assumed string-shape at 40mm above each nosing and 25mm below each step's underside corner. A flexi-curve lath of narrow hardboard strip was then held to the marks and the curved easings were drawn. Note that – because the underside of this stair was visible – the steps (as shown) were only housed in to their exact thicknesses; no tread- or riser-wedges were used. However, to compensate for this omission, each tread had two tenons which were through-mortised and wedged to both of the strings. The approximate positions of these 25mm-wide mortise-and-tenon connections are indicated on rods (b) and (c) above.

## Tread and riser details

*Figures 6.50(a)(b)*: The tapered treads were of 18mm-thick softwood exterior plywood, with 22 × 12mm redwood lipping glued on as nosings and shaped with an overhead router. The fretwork risers were also of similar plywood, but 9mm thick. These were grooved into the underside of the treads to a minimal depth of 4mm. As shown in Figure 6.50(a), the positioning of the groove created a recess to accommodate the 19 × 10mm scotia mould, between the riser-face and the backside of the lipping. To complete the construction, continuous lengths (rather than individual blocks) of 28 × 28mm cavetto-shaped mould were used as glue

**Figure 6.50 (b)** *Elevation of the simple fretwork pattern pierced through the riser-faces, emanating from a line of 64mm Ø (diameter) holes drilled with a holesaw. The radial triangular shapes were drilled for saw-entry and carefully formed with a jigsaw. This view also shows the step's nosing above the scotia mould and – at the base – the extended and pilot-holed riser-allowance for connection to the back-edge of the tread below.*

blocks to present a more visual finish to the underside of the stair. The fretwork risers were drilled and countersunk at approx 200mm centres on their lower faces for eventual gluing and screwing to the backside edges of the adjacent treads.

## Bullnose-ended step details

*Figures 6.51(a)(b)(c)(d)(e)*: The easy way to produce the segmental-rounded face-edges on the tapered, bullnose-ended tread was to use a solid timber board of 22mm thickness, as seen in Figure 6.51(a) – the face edges and end grain of which could easily be shaped to form the nosing. But, I chose to do it the hard way because 18mm softwood ply had been used

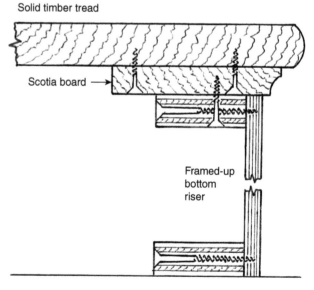

**Figure 6.51 (a)** By using a solid timber tread for the bottom, bullnose-end step, the additional work involved with a plywood tread requiring a lipped, gluelam nosing would have been avoided.

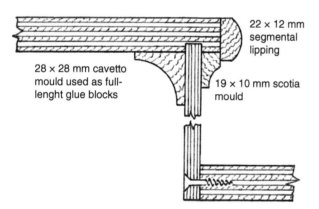

**Figure 6.50 (a)** *Vertical section of the tread-and-riser construction.*

Plywood tread with gluelam nosing

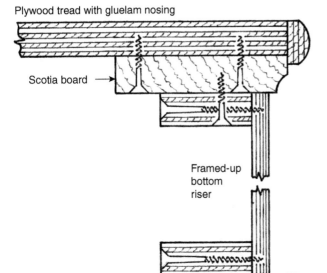

Scotia board →

Framed-up
bottom
riser

Figure 6.51 (b) A *section* through the actual bottom
step with its thicker scotia board and gluelam nosing; the
detailed illustrations of the construction of these – and the
open-boxed, pierced and bullnose-shaped plywood riser
are given below.

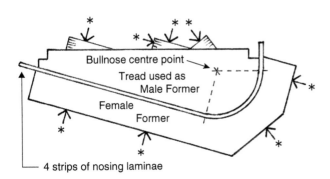

Bullnose centre point
Tread used as
Male Former
Female
Former
4 strips of nosing laminae

Figure 6.51 (c) A *plan view* of the bullnose tread used as
a male jig-former and a purpose-made female jig-former
made from 18mm plywood. Three temporary wedge-
shaped blocks were glued onto the tread's back edge to
provide four tangential sash-cramp positions (**). Note
that the 18mm ply female former was subsequently used
to make the hockey-stick shaped back-supports for the
open-boxed bullnose riser.

on the straight-edged tapered steps and I felt com-
pelled to use this material again. This was because
a non-carpeted stair was planned, with a varnished
pine-appearance and I did not want to see a change of
material on the treads, with a different pattern of grain
'grinning' through the clear varnish. The problem was
that by using 18mm plywood, I created an extra task
of bending the 22 × 12mm segmental-shaped lipped-
nosing around the bullnose end.

Figures 6.51(b) and (c) indicate how this was
done by using a technique referred to in Chapter 3

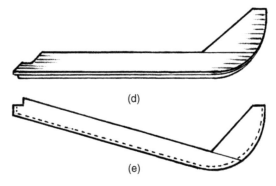

(d)

(e)

Figure 6.51 (d) A *3D view* of the made-up scotia-board;
and (e) A true **Plan View.**

as *gluelam* (glued and laminated) *construction.* Four
straight-grained strips of redwood were produced
measuring 23 × 3mm, which were glued and cramped
together around the step's outer edge, between a *female
jig-former* and the shaped edge of the step (acting as a
*male jig-former*). Note that the 1mm oversized width
of laminae was to allow for *cleaning-up* any slight
irregularities after gluing.

Before gluing the four laminae to the male-
former – not the female-former! – they were well
soaked with boiling water to increase their moisture
content and pliability, then (after an hour or so) they
were slowly cramped between the formers without
being glued and left in this state overnight. After
being released the next day and left to air for an hour
or so, wood glue was applied and they were cramped
up again and left until the following day. After trimming off the slightly oversize lengths, the top- and
underside-edges were cleaned up (to a 22mm finish)
and the segmental-shaped nosing was applied on a
fixed, overhead router.

Note that illustrations 6.51(a) and (b) show the
traditional technique for applying scotia moulds to
shape-ended steps – by applying a *stuck* mould to the
edges of shaped *scotia boards* that are then screwed to
the underside of the tread.

## Bullnose riser details

*Figures 6.52(a)(b)(c)(d)*: The approximate length of
this riser was gained by hooking up and flexing a tape
rule around the nosing-edge of the bullnose tread.
Its actual width (height) was 190mm rise minus the
tread-thickness (18mm) and the scotia-board thickness (19mm) = 190 – 37 = **153mm**. To set out the
fretwork pattern – which was based on a concept of
64mm Ø holes – the hole's diameter was deducted
from the 153mm riser-height, leaving 89mm. This was
divided by 4 to create two 22.25mm horizontal bands,
one at the top, one at the bottom, and a 22.25mm

Backside view of bullnose-end of bottom riser

(a)

Plan view of saw-kerf grooved riser

(b)

**Figures 6.52 (a)** and **(b)** *Part elevation and plan* showing the 15 saw-kerf grooves that were required in the backside of the bottom riser. Note that the un-kerfed end on the left was for the step's square entry into the front of the newel post.

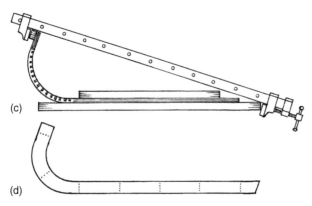

(c)

(d)

**Figure 6.52 (c)** As illustrated, the straight portion of the riser was temporarily supported by a 100mm-wide par board on each side, held by screws fixed through the open fretwork, then the 4 × 8mm timber inserts (staves) were positioned into the glued grooves and the end bent up and held with a sash cramp; and **(d)** the internal shape of the step was tested (during the cramping) with one of the two hockey-stick shaped and drilled back-supports previously prepared for the riser from 18mm plywood (shown in Figure 6.51(b)).

circular band around the 64mm Ø holes that accounted for the other two divisions. These bands created the radial triangles that were removed by the jigsaw. The challenge in setting out was in getting the rows of circles to finish with a full banded circle at each end of the varying riser lengths. And this was achieved by allowing the linkage bands between the circles (which had to be less than two bands, anyway, to achieve structural linkage) to overlap by a slightly greater or lesser amount on each riser.

The bottom riser required nine banded 64mm Ø holes and, after these had been drilled out and the

triangular fretwork completed, the bullnose end of the riser containing three of the holes had to be bent to shape. This further challenge was overcome by an experimental mixture of traditional techniques involving saw kerfs and staving. Vertical, saw-kerf grooves were cut across the backsides of the three ring-bands – and these 6mm-wide grooves were made slightly v-shaped in width. Their depth removed four of the five cross plies to expose a 4mm width of face ply. The idea was that this would act as a pliable veneer and allow the plywood to bend easily. As illustrated above, to strengthen the exposed portions of bent face-veneer (and reconsolidate the 9mm ply), 4mm × 8mm timber inserts (staves) were glued into the vertical grooves when cramping up.

After the usual glue-setting period, the slightly protruding staves were cleaned up on the concaved face of the riser and one of the 18mm ply, shaped back-supports was glued and screwed into its top position. The sash cramp was then released and the staves that were exposed in the openings of the fret-work were carefully removed with a fine, sabre-type keyhole saw. The half-framed riser was then glued and screwed to the underside of the scotia board (that had been previously glued and screwed to the underside of the bullnose tread) and – to complete the bottom-step assembly – the second hockey-stick shaped back-support was glued and screwed to the base of the riser.

## String details

*Figures 6.53(a)(b)(c)*: The three strings in this staircase are technically described as 1) the *inner (wall) string*; 2) the *outer (return) string*; and 3) the *outer (wreathed) string*. The first two were developed on rods, as shown in Figure 6.49(b) and (c) and – because of material availability – they were comprised of 30mm-thick plywood that I built-up by bonding 18mm- and 12mm-thick pieces together. Partly to conceal the edge appearance of the plywood and partly to create a wider surface for the balusters on the outer string, the top edges of both strings were lipped with 7mm-thick round-edged capping; the round edge was flush on the wall string, but projected by 7mm from each face of the outer string. The capping's minimal thickness of 7mm enabled it to bend to the curved easings of the two strings' top edges.

Regarding the *wreathed (coiled, spiral-shaped)* string, I chose at the outset to simplify this from that shown on my original drawing (Figure 6.47) and it became a simple floor-standing quadrant shape of solid, two-piece timber that was joined (by dowelling and gluing) to the extended, square sections of the adjacent newel posts.

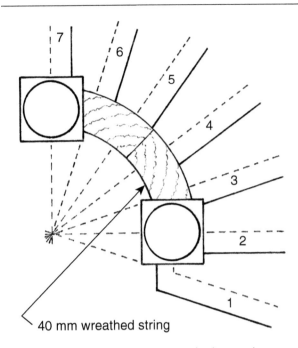

40 mm wreathed string

**Figure 6.53 (a)** The two-piece wreathed string, butt-jointed to the newel posts.

**Figure 6.53 (b)** The wreathed string was made in two parts for economy of material and, as shown, to minimize the shaped work – especially the concaved shape, which was achieved with a beech round-edge moulding plane and a curved cabinet scraper.

The above setting out, at 6.53(a), being more detailed than this part of the rod shown at 6.49(a), shows the broken lines of the riser-faces emanating at 18° from a centre point initially created as the corner of a right angle marked on the rod. From thereon, the position of the quadrant-shaped wreathed string – which determines the position of the two newels and can move towards the centre point or away from it – was trial-and-error tested with a compass until its inner face met and was slightly in excess of the 50mm minimum tread-width regulation. (Note that the actual stair has a minimum 'going' of 50mm, as stated in a previous AD K1 and, therefore, is now over-compensated with a minimum *tread-width* of 72mm). Having determined and marked the string's inner quadrant, the outer quadrant, representing the string's thickness of 40mm, was marked. This then enabled the 68 × 68mm square portions of the newels to be positioned and marked (in relation to the rule that the

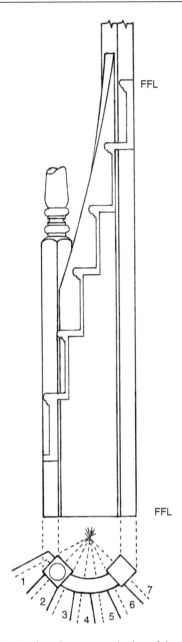

**Figure 6.53 (c)** The *elevation* and *plan* of the wreathed string and its newel posts, indicating that the unglued assembly was seated on the rod to allow the riser-faces to be marked and taken up to plot the steps in relation to the known riser-heights.

riser-face equals the centre of the newel). Finally, the nosing edges were marked at 22mm from the risers.

Note that a centre-point is normally used to plot the *nosing edges* of radiating tapered steps, not the *riser faces* as above; however, when a wreathed-string and a wreathed-handrail are involved, the geometry of the situation changes because they share a common centre with the radiating riser faces – not the nosing edges. Technically, when tapered nosing edges do not share a common centre, as above, they are referred to as *dancing steps*. Normally, it is the riser-faces (by virtue of being parallel to the nosings) that do not share a common centre.

Once the nosings were marked (including the scotia moulds, which were to be housed in as well), a 20mm helical margin line was established above them with the aid of a narrow hardboard flexi rule. Then the ends of this line were squared across the string's narrow edges to produce a high and low mark on the face-side edges of the concaved side. The flexi rule was applied again to pick up these marks and the second helix was drawn to complete the marking out of the string's wreathed edge. The excessive (but reusable) waste above the helical edge was cut away with a reciprocating-type sabre saw – after the housing out of the steps, but obviously before gluing the dowelled edges of the string to the newels. Throughout the cutting operation, careful vigilance had to be kept on both sides of the twisting helical lines. Then the edge was finished to the lines with a flat-bottomed spokeshave.

## Tapered-step details

*Figures 6.54(a)(b)(c)(d)(e)*: As previously mentioned, oversized tapered treads are laid on the rod and carefully squared up to the rod's nosing lines. Then the extremities of each different shape (i.e. *square winder, skew winder, kite winder*) are squared up and marked out on the face sides. Extended parallel lines are then marked, equal to the depth of the housings (normally 12mm), to determine the final cut-lines. However, if a tapered-stair arrangement – unlike the one shown here – is to be fitted in a corner, further allowances of at least 10mm should be added beyond the cut-off lines to compensate for any non-squareness of walls that the site-fitter may encounter. As illustrated below, I only had to allow extra beyond the housing depth for the projecting tread-tenons that were to be mortised through the two plywood strings.

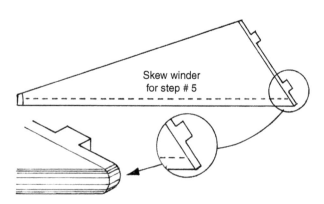

Figure 6.54 (a) The 90° chisel-pared, eased-nosing of acute-angled tapered tread # 5.

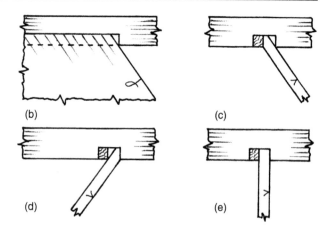

Figure 6.54 (b) The non-eased nosing of obtuse-angled tapered tread # 4 in relation to its square-ended, nosing-shaped housing; (c) the string-housing and end-treatment of the riser for step # 4 (similar to that required for step # 3); (d) the housing- and riser-treatment for the acute riser entry for step # 5 (similar to that required for step # 6; and (e) a square riser-entry, not used here, but shown for comparison. Note that these connections to the stair-string housings are usually wedged, as shown, but – as mentioned previously – because this stair was exposed on the underside, the step-housings were not given allowances for wedges.

Further considerations concern minor easings that have to be made to the nosing-ends of some of the tapered treads and to the face-side and backside ends of risers at the points of entry into the housings. This 'easing' has to be made where nosings and risers (and scotia moulds, if used) are at acute or obtuse angles to the string, as illustrated above. For practical reasons, the square-sided housings cannot be undercut to accommodate these ends.

## Newel post details

*Figures 6.55(a)(b)(c)*: The lathe-turned portion of the four newels (and two half-newels for the abutment to the walls) were finished with projecting spigots at each end that were eventually glued into corresponding holes in the square-shaped newels below and into the handrail and handrail-caps above. The half-newels were made by cramping two half sections together, with a sheet of glued lining paper sandwiched between them. After being turned, the end-grain glue-line was easily levered apart with a firmer chisel, to produce the half newels.

The square newels had a finished size of 69 × 69mm and their height was governed by a marginal amount above the wreathed string, as can be seen in Figure 6.53(c). The segmental-rounded top edges seen in 6.53(c) and below in Figure 6.55(c), to the bottom edges

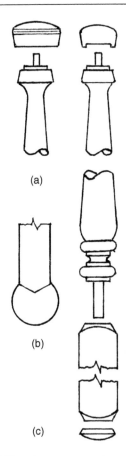

(a)

(b)

(c)

Figure 6.55 (a) This shows a newel cap suspended above its position on the spigot-end of one of the lathe-turned newel posts; (b) shows a plan view of the newel cap with the 68 × 45mm handrail mitred to it with bisected angles; and (c) shows the four separated components of the newel post in the central position of the galleried balustrade, i.e. handrail (grooved for baluster-ends and spacer fillets); spigot-ended and turned newel post; the square portion of the newel post; and the pendant cap. Note that, of the two pendant caps required, one was cut in half for the half-newels.

as well, were easily marked with a round drawing aid (such as a paint-tin lid) and formed with a smoothing plane, rasp and an abrasive sanding block. Four newels and two half-newels were required – and the square portion of the half newel on the stair was housed out to carry the end of the inner (wall) string and mortised to receive the tenons of the return outer string. Finally, the square-ended newels were finished with slim, lathe-turned pendant caps glued to their underside.

## Apron lining details

*Figure 6.56*: Because the 9mm-ply apron lining was pierced with an open tracery pattern, I covered the rough face-side of the sawn landing-trimmer with a 6mm plywood lining. And to gain a better visual effect

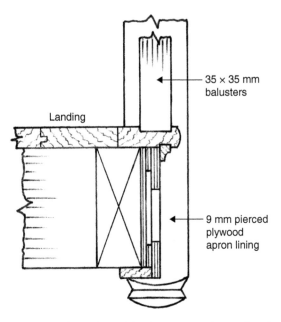

Landing

35 × 35 mm balusters

9 mm pierced plywood apron lining

Figure 6.56 The galleried landing's apron-lining details related to the central newel.

by creating an open background, continuous narrow strips of 6mm ply were then fixed to the top and bottom edges, as illustrated, to which the apron lining was fixed. The landing nosing-piece was grooved to receive the scotia mould and it was also housed to a 6mm depth to receive the ends of the skew-nailed balusters (spindles). The controlling datum for the apron lining's lateral position was the centralizing of the spindles to the newels.

## The curved handrail over the outer return-string

*Figure 6.57*: With such a shallow curve and slightly ramped-end, this handrail could have been made three

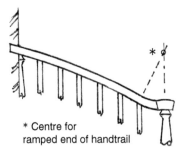

* ∅

* Centre for ramped end of handtrail

Figure 6.57 The curved handrail follows the shape of the curved string-top below and is mitred into the newel cap after being eased into a level position by the handrail ramp. As shown, the end of the handrail above the half-newel was sunk into the interior wall by 40 to 50mm.

ways; 1) formed with 3 pieces of 15mm horizontal gluelam laminae in a jig-former (with a short-length top laminae added after for the ramp); 2) formed by G-cramped, gluelam construction with 3 pieces of pre-shaped *vertical* laminae of 23mm, or 4 pieces of 17mm laminae; or 3) cut to shape from one piece of 100 × 75mm timber. This third method was used – and after surfacing and thicknessing the timber to the handrail-width of 68mm and marking out the handrail shape with the aid of the pre-curved string, it was cut on a bandsaw and brought to a finished shape with a compass plane and spokeshave.

## Wreathed handrail over the wreathed outer string

*Figure 6.58*(a): It has to be said that – because of the close proximity of the newels on each side of the wreathed string – this stair design presented a challenging situation for the wreathed handrail to rise up from the level newel-cap at the bottom and be jointed to the newel cap of the adjacent newel at the top. It became apparent that about two-thirds of the upper wreathed-handrail would be almost vertical and look unattractive so close to the left-hand newel. So, contrary to the original plan, the wreath was modified and made to finish at the segmental-curved top of the square portion of the receiving newel post. This looked

more attractive and when one's hand moved up from the handrail, it felt natural to start holding on to the turned portion of the newel just above it – as demonstrated in Figure 6.48(b)'s photo.

## Setting out the wreath

*Figure 6.58(b)*: The key to this is the making of a 6mm plywood template called a *face mould* and the geometry for the wreath – to make it more understandable – is given after a view of the three-dimensional template below. The two central lines shown on the face mould are tangents to the wreath's inclined centre and the other two lines, one at each end, are extensions for jointing tolerances.

*Figure 6.58(c)*: The imaginary quadrant-shaped centre line of the wreathed string in Figure 6.53(c), between the two newels and riser-faces 2 to 7, is the basic information that was used to develop the wreathed handrail above it. And to help visualize how this inclined single-twist wreath is transformed from geometric principles to a hand-manufactured item, the quadrant-shaped centre line in Figure 6.58(c) below – described from *a-b* to *d* – has been drawn within a

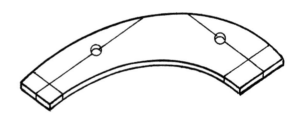

Figure 6.58 (b) The 6mm ply face-mould template. Note that the holes (of 18mm Ø) are for easier viewing/location of the tangent lines to be marked on the handrail wreath.

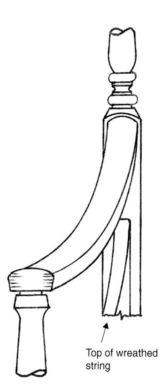

Figure 6.58 (a) The handrail wreath in relation to the adjoining newel posts.

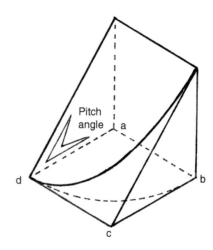

Figure 6.58 (c) The geometry of the quarter ellipse above the square-prism quadrant.

square prism (*a, b, c, d*) with its top surface inclined
to the known stair-pitch angle. If you can imagine
the true, marked quadrant-shape of the prism being
sawn around vertically with a bandsaw, the shape that
that would produce on the inclined top-surface would
be a true quarter ellipse (as drawn here), equal to the
inclined centre-line of the wreathed handrail. And this
quarter-ellipse is also equal to the un-drawn centre-
line of the above face-mould template.

*Figure 6.58(d)*: To set out the handrail wreath,
reference was made to the original rod and the plan
view of the wreathed string, (Figure 6.53(c)). The
centre line of the string was drawn, as illustrated,
with the required quadrant-shaped width of the
handrail added. The central quadrant was then boxed
in with a plan view of the square prism marked *a, b,
c, d*. From centre *a*, a quadrant was struck to the left
to end on an extended line from *b*. With *b* as centre,
a semi-quadrant was struck from *d* to the same
extended line. As shown, the two left-hand lines
were extended up to a so-called X-Y line separating HP (the horizontal plane) from VP (the vertical
plane) and the other two were extended up further
into the vertical plane. The stair-pitch line (C) was
drawn on the left until it struck the line extended up
from *a*, to create the bevel required for the wreath.
From this point, horizontal line (B) was formed. Line
(A) was then drawn to create the important trapezoi-
dal shape which provided the three lengths (at A, B,
C) of the axes for the quarter ellipse required to draw
the face mould.

*Figure 6.58(e)*: With reference to Figure 6.58(d)'s
trapezoid, to develop the face-mould template's shape,
the length of inclined line (A) was measured and
drawn horizontally in the development below. Next, a
compass was set up to represent the length of inclined
line (C) and arcs were struck above and below line (A);
first, above from the left extremity of line (A), then,
below from the right extremity. A similar technique
was used for trapezoid line (B) and, again, arcs were
struck above and below line (A), but from the opposite
ends to intersect with the arcs struck previously. Lines
B¹ C ¹ B ² C ² were drawn carefully through the arcs
to form the rectangular prism. Lines B¹ and C¹ were
the axes for forming the two quarter-ellipses of the
face mould by the short-trammel method (illustrated
in Chapter 8: Geometry for Curved Joinery). Lines B²
C ² are marked on the face mould to identify the position of the wreath's central tangents.

## Making the twisted wreath

*Figure 6.59*: The piece of thick (or built-up) timber
for the wreath should be straight-grained, knot-free
and, obviously, of the same species as the adjoining

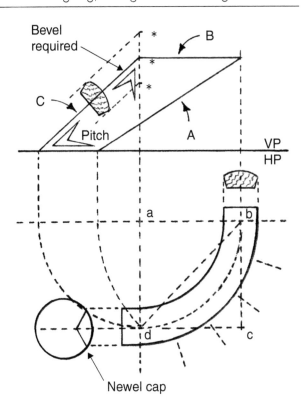

Figure 6.58 (d) The geometrical development of the
widest width of the face-mould template at the three asterisks and the lengths of the face-mould axes at A, B, C.

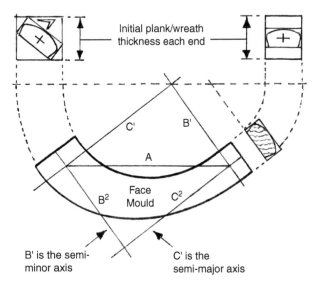

Figure 6.58 (e) The development of the face-mould template and – as seen in the top left corner – the development of the wreath's initial plank thickness by applying the
plumb bevel through the drawn handrail's centre line.

handrails. The face-mould template is first used to
mark the basic wreath-shape on the plank's planed
face-side – and it should be laid diagonally so that
the unavoidable short grain is equally distributed and

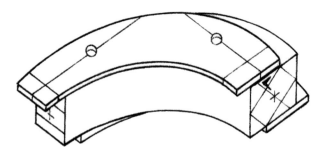

Figure 6.59 The handrail wreath being marked on its top and bottom faces with the face-mould template.

thereby lengthened. Then the wreath is cut carefully to the double-curvature shape, ideally on a narrow bandsaw machine and its sides are planed to a fair finish with spokeshaves and/or a sanding bobbin and tabled disc-sander.

Next, the two tangent lines (previously transferred from the face mould) are squared down and marked on the ends of the wreath, as illustrated in Figure 6.59. These small crosses pinpoint the centre-lines of the level handrail at the narrow end and the twisted handrail at the wreath's wide end. And they control the pencilled outline of the plain,

rectangular shapes of the handrail to be drawn at each end. However, note that the centre of the twisted handrail has to be altered to suit the stair's plumb bevel.

When marked on both faces, as illustrated, the wreath was fixed firmly to a wedge-shaped block that had been made to the stair-pitch angle; (countersunk screw-holes had been drilled in the block's underside, so that two fixings could be made to the underside of the wreath). With the wreath now positioned in its correct vertical plane, the marked outer-edges were cut by pivoting the wide wedge-shaped block very carefully around the narrow bandsaw.

After cleaning up the edges with spokeshave planes, the so-called *falling lines* (helixes) were pencilled in on each side – to depict the top and bottom curvature of the wreath – and these twisting shapes were removed with a traditional bow saw. This is a challenging task that requires constant observation on the two separated and twisting lines. Again, the sawn surfaces were cleaned up with spokeshaves. Finally, when satisfied with the basic rectangular shaped wreath (and not before), finger-gauged pencil lines are made and the actual handrail shape is applied – again with the aid of spokeshaves and cabinet scrapers, etc.

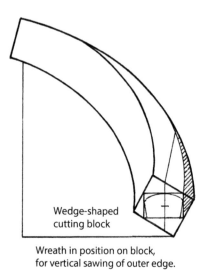

Wedge-shaped cutting block

Wreath in position on block, for vertical sawing of outer edge.

Figure 6.60 (a) Wreath in position on block, for vertical sawing of outer edge

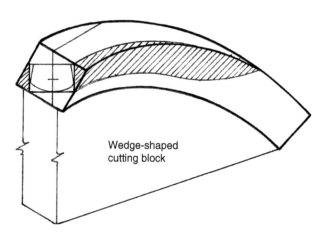

Wedge-shaped cutting block

Figure 6.60 (b) Wreath upside down on block, for vertical sawing of inner edge

# 7
# Designing and making shelving arrangements

## INTRODUCTION

Although designing is not the designated domain of a joiner, a love of the craft and an acquired understanding of wood and wood-derived products can inspire many of us to create our own designs. However, even if our designs are not revolutionary, we do need to consider shelf thicknesses in relation to the weight of the books or other items to be placed upon them; and in so doing, we become involved with the principles of structural mechanics. This is because the shelves, usually required aesthetically to be kept thin in depth in relation to their span, are likely to bend (deflect) in their middle area under their load. This is rarely of concern regarding the actual shelves reaching a point of collapse, but it must be realized that bending is the most severe form of stress in a shelf. And to avoid the unacceptable appearance of excessive deflection, we can work out a shelf's required mechanical thickness by applying simple or complex mathematical formulae – in a similar way to working out the depth of timber beams or floor joists. Not being a structural engineer myself, but having designed and built many successful shelf arrangements over the years, the following notes and illustrations are based on practical intelligence and common-sense mechanics related to *simple* formulas.

## SHELF MATERIAL

Although this chapter will refer to engineered-wood products such as plywood and blockboard, and wood-derived products such as chipboard and MDF, etc, which could be used for shelves, the calculations given here for the critical shelf-thickness (depth) are mainly based on the use of good quality, solid redwood or a hardwood species of at least equal density and strength.

## BEAMS OR JOISTS LIKENED TO SHELVES

Because the structural stability of suspended wooden beams or joists is mostly dependant on their cross-sectional depth, suspended shelves – by comparison – are like ignorantly-positioned, suspended wooden beams of rectangular cross-section, lying wrongly on their widest sides – and yet they have to support so-called *live loads* (be it only of books, etc) similar to floor joists that carry a *live load* of people and furniture, etc. As beams or joists, therefore, shelves are structurally weak and – to avoid excessive deflection under load – must be limited in span in relation to their thickness and their intended load. All relatively thin, loaded wooden shelves will suffer a downwards deflection to some extent, but, with calculated design, this deflection should not exceed (in my opinion) 1 in 300 (i.e. a shelf with a 900mm span should not have more than a 3mm mid-area deflection under load).

## UNDERSTANDING BASIC MECHANICS

*Figures 7.1(a)(b)(c)*: The fibres of loaded beams or shelves, as illustrated below at 1(a) and (b), are subject to a crushing effect referred to as *compressive stress (compression)* across their horizontal width below the topside – and to being torn apart (stretched) by *tensile*

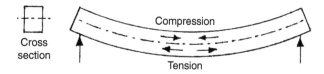

Figure 7.1 (a) Exaggerated deflection of a simply supported beam, with arrowed indication of the compressive and tensile stresses; the broken line through the centre represents the neutral layer. The extreme vertical arrows depict the bearing points.

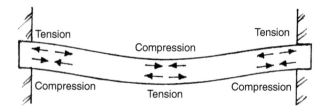

**Figure 7.1 (b)** Exaggerated serpentine-shape deflection highlighting the changes in the compressive and tensile stresses on a beam with built-in end-bearings.

**Figure 7.1 (c)** A simply supported, overloaded beam displaying exaggerated deflection and a theoretical impression of the effects of *horizontal shear.*

*stress (tension)* across their horizontal width above the underside. Being at maximum stress on the top and bottom outer-surfaces, where these two theoretical forces meet in the horizontal middle area of a beam or shelf, they neutralize and have no stress. This area is referred to as the *neutral layer*, or *neutral axis.* Because the compressive stress in the layer of wood-fibres above the neutral axis causes a limiting- or balancing-effect on the tensile stress in the layer of wood-fibres below the neutral axis, it follows that the greater the thickness of the wood-fibres above and below the neutral layer, the less bending (deflection) will occur – even though the weight of the beam or shelf will increase.

Another way to appreciate the validity of the limiting- or balancing-effect of the opposing compression and tension forces in a rectangular beam is to study the purposely-exaggerated concaved beam-shape at 7.1(a) and question its ability to actually (physically) compress and *shorten* its length on top and stretch and *expand* its length on the bottom (as, geometrically, any such concentric curvature demands) without fracturing (rupturing) the beam on the underside. In theory, of course, an excessive dead load on a *simply-supported beam* (the ends of the beam resting on supports, but not built-in) can also cause *horizontal shear*, as illustrated at 7.1(c). This causes a separation of the beam's horizontal fibres (to a greater or lesser extent), which tend to slide over each other as the top of the beam is shortened and the bottom lengthened by bending.

Understanding these simplified explanations of the theory of basic beam-mechanics should help in visualizing its modifying effect on the different shelf-supports described below. The supports are also

referred to as *bearings* and – depending upon which of the three usual kinds are used – it has to be realized that each of them affect the degree of *stiffness* or *elasticity* of the shelf and its ability to resist deflection under load – as mentioned above.

## SHELF SUPPORTS

### Built-in end-bearings

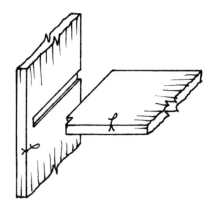

**Figure 7.2 (a)** Part of a side cheek showing a stopped-housing and part of a shelf showing the recess-shouldered front-end positioned for assembly.

*Figure 7.2(a)*: There are traditionally three different ways to support wooden shelves – apart from modern innovations by manufacturers and individual designers. The first, illustrated at 7.2(a), is where the ends of shelves are housed (usually stop-housed) into the cheeks (sides) of a shelf unit by one-third the cheek's thickness – the housing-stop distance equals the thickness of the shelf away from the edge. Note that if this is done neatly – i.e. the shelves fit tightly in the housings – you will achieve more mechanical support by having altered the tensile and compressive stresses, as indicated in Figure 7.1(b) above. Another advantage is that the restraint created by the housings will restrict warping or cupping of the shelves.

### Unfixed end-bearings

*Figures 7.2(b)(c)(d)*: The second traditional support, illustrated at 7.2(b), is where the shelves rest on – or can be fixed to – wooden end-bearers. However, it must be realized that shelves *simply-supported* in this way (whether fixed or not) are entirely or mostly subject to full-span stresses of compression and tension, as described above and indicated in Figure 7.1(a).

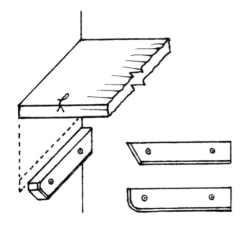

Figure 7.2 (b) Part of a simply supported shelf shown in a raised position to allow a view of the 45° splay-chamfered bearer fixed to the wall; two alternative end-shapes to edge-chamfered bearers are shown alongside.

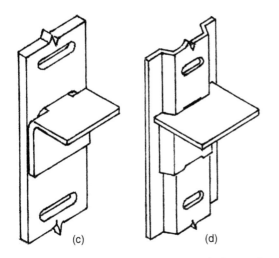

(c)                    (d)

Figure 7.2 (c) Part isometric-view of a Tonk Flat Bookcase Strip and (d) a part view of a Tonk Raised Bookcase Strip; the clips shown are a heavy-duty type.

Another type of bearing, illustrated above at (c) and (d), for simply-supported shelves is available in the form of vertical metal-strips that are pierced with closely-spaced, small slots. The slots support small, flat-lying metal lugs (referred to as *clips*) that act as interchangeable bracket-bearers, giving infinitely variable shelf-spacing options. The vertical strips – two normally required at each end of the shelves – are either of a shallow, inverted u-shaped surface-fixing type, or are comprised of flat strips of metal that require housing into double-grooved, vertical channels to bring them flush with the surface of the bookcase cheeks. Note that the stepped inner groove accommodates the bracket-ends of the

shelf-supporting clips. The flush-finishing type is more professional-looking – especially when used with hardwood.

An online search for these so-called *Tonk strips* at www.ironmongerydirect.com revealed that these traditional items of ironmongery are still available in both types and are described as *Tonk Flat Bookcase Strip* in lengths of 1829 × 19mm and *Tonk Raised Bookcase Strip* in lengths of 1829 × 24mm. They are available in a variety of finishes, namely zinc-plated, brown, electro brass, satin-nickel plated, polished solid brass and polished chrome (on brass). Note that the last two types are not available in the Raised Strip range. Four clips are required per shelf – and a specially-designed router cutter is available for forming the stepped, vertical grooves in the cheeks.

## Indented and mid-area bearings

*Figure 7.2(e)*: The third traditional form of support – used predominantly for storage shelves of short or long lengths – is right-angled brackets. Although these can be made of wood, they need a diagonal brace and – unless the brace is ornate – they tend to look bulky and ugly. Also, making small, wooden brackets is tedious and time-consuming. Therefore the cantilever-type metal brackets – especially the so-called *London-pattern* type, illustrated below – are quite commonly used. They can be screwed directly onto the wall, via their three fixing holes (two at the top, one at the bottom), but it is more practical for them to be fixed onto pre-positioned, plugged-and-screwed vertical wooden cleats. The cleats, usually made from ex 50 × 25mm softwood, should be neatly chamfered on the two face-side edges and bottom edge.

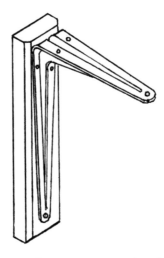

Figure 7.2 (e) London pattern, pressed-steel bracket fixed to vertical wooden cleat.

## Spacing of indented end-cleats and brackets

Regardless of whether the ends of these shelves butt up against walls or not, the end-cleats and brackets are normally set in (indented) from the shelves' ends by a certain amount, thereby creating cantilevered ends. And if the amount of indent is calculated, rather than guessed at, the strength of the remaining shelf-lengths between the brackets will be significantly improved – as will the degree of deflection. The simple formula for a *uniformly distributed load* is:

Indent = 0.207 × Overall Length of shelf.

For example, a shelf with an overall length of 1500mm would be 0.207 × 1500 = 310.5 (say 310mm) indented bracket each end. Note, though, that the formula is for a theoretical *uniformly* distributed load – and yet in practical terms, the shelves are more likely to be *unevenly* loaded. I would recommend, therefore, that the formula for *unevenly distributed loads* should produce an *indent* reduced by about 25%. This would produce a formula of:

Indent = 0.155 × Overall Length of shelf.

Therefore, the same length shelf of 1500mm would be 0.155 × 1500 = 232.5 (say 232mm) indented bracket each end – instead of 310mm for shelves with a uniformly distributed load. Note again though that the remaining 1036mm between the brackets should be checked out against the shelf-span formula below, as certain lengths of shelves will obviously require intermediate brackets.

## Back-edge bearers

*Figure 7.2(f)*: Note that shelves being supported by vertical cleats and brackets can be additionally strengthened by pre-fixing horizontal shelf-length bearers to the wall, directly below the shelves and seated on the cleat-tops, as illustrated below. The back-edge bearers would be of the same section and chamfered detail as the cleats

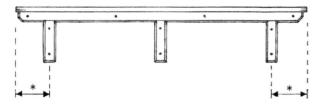

Figure 7.2 (f) Front elevation of a mid-area shelf fixed to a back-edge bearer seated on vertical cleats. The three brackets are omitted. (*) The asterisks highlight the calculated indents for the brackets at each end. Note that – because of the chamfer – the cleats should be scribed or housed-in to the underside of the bearer.

and the shelves should be nailed or screwed to them – especially if using relatively weak shelving material, such as chipboard, Contiboard, OSB (oriented strand board) or MDF (medium-density fibreboard).

## Splayed bearings to shelf-ends and sides

*Figures 7.2(g)(h)*: The traditional method for joining solid- or slatted-shelves at right-angles to each other – as in a small storage cupboard or (in the case of slatted shelves) in an airing cupboard – is to splay the ends to an angle of about 60° to 70° and form splayed housings in the shelf-sides to accommodate them. These splayed joints (or bearings) are usually dry-fixed and panel-pinned, but could be glued as well.

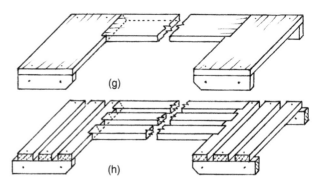

Figure 7.2 (g) Three-sided shelf arrangement in a cupboard, with a splay-ended shelf between the outer shelves; note that the back-edge bearer (only partly shown) should be taken right across the wall, under the three shelves. (h) Slatted shelves.

# LOADS ON SHELVES LIKENED TO LOADS ON FLOOR JOISTS

The so-called *live-loads* on floor joists in residential properties refer to the *aggregate* weight of furniture, fittings, etc, and the number of people who are likely to use the dwelling – and the formulae to determine the sectional size of the joists required over certain spans incorporate tested margins of safety and do not have to take into account the *actual* weight of the furniture or fittings, the *actual* number of occupants and visitors likely to load the floors, or their *actual* en masse bodyweight related to anorexia or obesity. Neither, therefore, does the formula given below for *residential (non-commercial or industrial)* shelf-thicknesses related to span have to take into account the *actual* size and weight of the books, magazines and other items to be placed upon them.

First, though, having likened the shelves to floor joists and because the formula for the shelves evolved from the joist-formula, I give below an example of the simple rule-of-thumb formula that exists for *depth of floor-joists related to span*. The formula was traditionally expressed in feet and inches, but was metricated by technical authors in the 1970s.

## Depth of floor-joists related to span

The formula for this is easily remembered as: *span over two, plus two*, or:

$$\frac{\text{Span}}{2} + 2$$

This is expressed literally as: Depth of joists in centimetres equals span in decimetres divided by two, plus two. Or, as an equation as:

$$\text{Depth of joists in centimetres} = \frac{\text{span in decimetres}}{2} + 2$$

For example, with a joist-span of 4 metres (40 decimetres), the equation would be:

$$\text{Depth of joists in centimetres} = \frac{40}{2} + 2 = 22\text{cm (220mm)}$$

Note that The Building Regulations' Approved Document A1/2 Floor Joist Tables for a small house recommends *195mm* × 50mm C16 structurally graded joists for a 4m clear span, which highlights a 25mm difference between the more precise Joist-Tables and the rule-of-thumb formula. This is because the latter has to slightly overcompensate to make up for its mathematical elementariness.

## Thickness of shelves related to span

The formula for this is easily remembered as: *span over fifty, plus two*, or:

$$\frac{\text{Span}}{50} + 2$$

This is expressed literally as: Thickness of shelves in millimetres equals span in millimetres divided by fifty, plus two. Or, as an equation as:

$$\text{Thickness of shelf in millimetres} = \frac{\text{span in millimetres}}{50} + 2$$

For example, with a shelf-span of 900mm, the equation would be:

$$\text{Thickness of shelf} = \frac{900}{50} + 2 = 20\text{mm}$$

Two further examples, with shelf-spans of 1200mm and 450mm would be:

$$\text{Thickness of shelf} = \frac{1200}{50} + 2 = 26\text{mm}$$

$$\text{Thickness of shelf} = \frac{450}{50} + 2 = 11\text{mm}$$

Note that calculated shelf-thicknesses usually have to be slightly increased or decreased by a millimetre or two to suit commercial timber sizes and manufactured-board thicknesses; but if a heavy load of large hardback books or glossy magazines and the like are anticipated for the shelves, decreases should be avoided.

# BLOCKBOARD OR PLYWOOD AS SHELVES

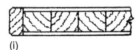 

(i)                                    (j)

**Figure 7.2** (i) 9mm-thick, lipped face-edge of 18mm-thick blockboard shelf; and (j) Similar lipping bonded to face-edge of 18mm-thick plywood shelf. Note the sharp arrises removed from the lipped edges.

*Figures 7.2(i)(j)*: Although I devised the shelf formula for good quality redwood and hardwood shelves, I find that it also works well on such material as blockboard (when the core strips run parallel to the span) and plywood (when the face grain of the outer plies runs parallel to the span). Also, these materials have the advantage of being more stable and will not suffer from warping or cupping. Note, though, that as illustrated at (i) and (j), shelves made from these materials need to be *lipped* on their face-edges to improve their finished appearance.

## MID-AREA SHELF-SUPPORTS

If a shelf or an arrangement of shelves has multiple bearing points or brackets, i.e. one or more inner supports in addition to a support at each end, as in Figure 7.2(f) above and Figure 7.3(b) below, the shelving will take slightly more load. This is because of the cantilever affect acting on each side of the mid-area bearing points, related to compressive and tensile stresses explained at the start of the chapter. Alternatively, in these situations, the shelving could be of slightly less thickness than the calculated thickness determined by the actual span between each of the multiple supports. These strength-considerations also apply to bookcase shelves when their rear edges are held by pinning or screwing through built-in back panels.

## LADDER-FRAME SHELF UNIT

Design was mentioned at the beginning of this chapter and is now the subject at the end of it. The illustration below shows part of a simply constructed ladder-frame shelf unit that I first designed and made over three decades ago. Similar units made in recent years have been only slightly modified.

### Original concept

*Figure 7.3(a)*: Wall-to-wall shelving was needed in the recesses on each side of my family's living room chimney-breast, above newly built-in storage cupboards of about 900mm height. Aesthetically, shelves supported by wooden end-bearers or indented brackets were not an option – and the other extreme of using solid bookcase-type cheeks was not quite what was wanted either. So, with simplicity and economy in mind, I thought of a ladder-frame design that could be fixed to each side-wall so that the *rungs* could support the shelf-ends.

### Frame material

The ladder-frame uprights are made from ex 25 × 25mm (usually reduced by timber merchants to a 20 × 20mm par finish) redwood or hardwood, which must be fairly straight-grained, straight and knot-free. If either one or both of the end ladder-frames are not up against a return wall, short lengths of the frame material are also required for modifications to the shelf-ends and rear-edges, as detailed below. These modifications create more connectivity between the fixed frames and the unfixed shelves. The horizontal

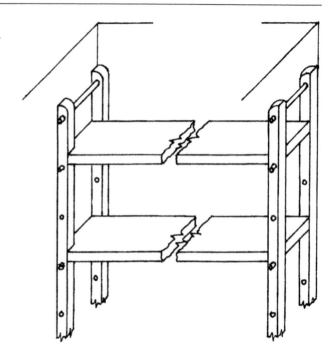

Figure 7.3 (a) Original concept of adjustable wall-to-wall shelving. Over the years this design – still using the same sectional-sized timber – has evolved into fixed, floor-standing units of varying heights (mostly floor-to-near-ceiling or coving), usually with one or two mid-area ladder-frames, depending on the overall unit width required in relation to the span calculations.

rungs are made from 9 or 10mm Ø (diameter) dowel rod – which is commercially available in redwood or light-coloured hardwood in standard lengths of 2.4m.

### Construction of the ladder-frames

*Figures 7.3(a)(b)(c)(d)(e)(f)(g)*: As illustrated, each pair of uprights are drilled right through their faces with 9 or 10mm Ø dowel-holes, equally spaced at 75mm centres throughout the entire height from bottom- to top-shelf dowel-bearers. Prior to drilling, though, the bottom parts of the uprights against the wall have to be scribed or modified to accommodate the projecting skirting member; and the top parts can be prepared for alternative finishes at the same time. Both of these items are detailed separately in Figures 7.3(c) to (g).

Each pair of uprights becomes a complete frame by inserting three end-prepared, projecting dowel-rungs in position and fixing them through the frame-sides with 18mm sherardized panel pins. The three positions for the fixed dowels are the extreme top, the extreme bottom and a mid- or near mid-area location.

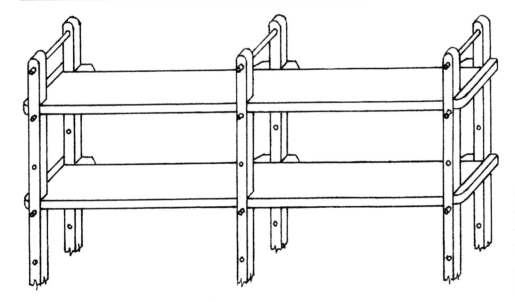

Figure 7.3 (b) The upper part of a floor-standing, open-ended shelf unit with one mid-area ladder-frame. (No additional dowel-rungs shown).

## Dowel-rung details

The length of all dowel-rungs equals the overall width of a frame (dependant upon the shelf-width decided upon), plus a small front dowel-projection of 6mm. After cutting the dowels to length with a fine saw, *all* ends must be rounded or chamfered to a good visual finish. This can be done with a rasp and glasspaper or – if available – by rotating each dowel-end safely on the bed of a fixed disc-sander machine. The essential number of rungs required is one for every shelf bearing – the idea being that if one or more of the shelves ever require repositioning, the rungs can be slid out and replaced in a lower or higher position. And the maximum optional number of rungs is one for every pair of frame-holes – thereby using them as a visual feature. I usually allow at least one additional rung per frame, per shelf, as – with floor-standing, open-ended frames – these retain and support the end books.

## Skirting allowances

*Figures 7.3(c)(d)*: As illustrated below, the ex 25 × 25mm frame member against the wall will usually require modifying to avoid clashing with a projecting skirting board. If the skirting is of a modern, simple design – usually with a 16 × 70mm finish – the base of the upright can be scribed to override it, as at (c), or it can be stepped over with a short, additional piece of framing glued to the upright, as at (d).

## Alternative finish to frame-tops

*Figures 7.3(e)(f)(g)*: One of the simplest ways to finish the frame-tops is with a semi-circular shape

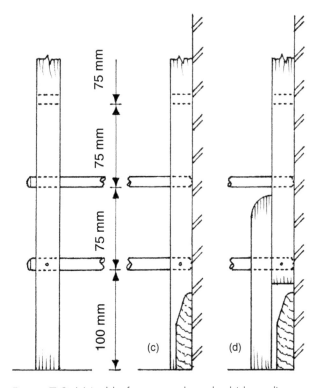

Figure 7.3 (c) Ladder-frame upright scribed (shaped) to override the skirting; and (d) the alternative treatment of modifying the upright to step over the skirting. Note the pinning of the bottom dowel-rung.

on the end of each upright, as at (e), with the fixed dowel-rung centred at 20mm below. Alternatively, a piece of framing can be used as a top member, bridle-jointed and glued to both uprights, as at (f). With this finish, instead of three fixed dowel-rungs, only two are required. Finally – although other ideas could be

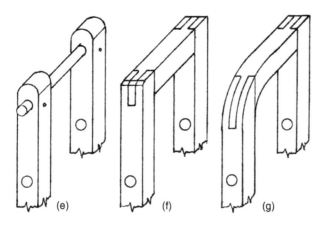

Figures 7.3 (e)(f)(g) Alternative top-members fixed to ladder-frame uprights.

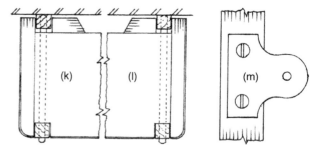

Figure 7.3 (k) and (l) Horizontal section showing bullnose-rounded side fillets on both ends of a framed shelf, as an alternative to the 45° bevel-ended fillets shown at (h) and (j) above; (m) Part rear-elevation of a ladder-frame upright and a brass picture/mirror plate (facing into the shelves) for wall-fixings; three plates per upright are recommended.

used – a good visual effect can be achieved by bridle-jointing a quadrant-shaped diagonal to each upright, as at (g). Again, only two fixed dowel-rungs per frame are required.

## Shelf details

*Figures 7.3(h)(i)(j)(k)(l)(m)*: As previously mentioned, if the end-frames are not fixed to return walls, more connectivity has to be created between the frames and the shelves. As illustrated below at (h) to (k), this is achieved by gluing and pinning (or screwing) shelf-thickness fillets to the end- and rear-edges of the shelves. Note that the bottom shelf requires extra notched recesses, as at (j), if the stepped framing at 7.3(d) is used. All notched recesses should be 0.5 to 1mm wider than the 20mm wide frame uprights to allow easy positioning and – if required – repositioning of shelves.

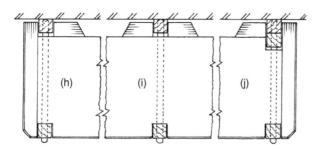

Figures 7.3 (h)(i)(j) Horizontal section through a ladder-framed unit showing: (h) Common shelf-end detail; (i) Common mid-shelf detail; (j) Extra recessed-notch detail to bottom shelf wherever the frame is stepped over the skirting board.

Figure 7.4 (a) Part of a newly built bookcase.

Figure 7.4 (b) Close-up of the stepped frame over the skirting and the additional notched recess in the rear edge of the bottom shelf.

Figure 7.4 (d) An end-view of a book-laden shelf arrangement comprising three ladder-frames.

Figure 7.4 (c) One of the picture/mirror-plate fixings facing into the shelves.

Figure 7.4 (e) A wall-to-wall softwood-stained and sealed book-laden shelf unit (designed and made in 1994), displaying quadrant-shaped frame-tops, as detailed at 7.3(g) above.

# 8

# Geometry for curved joinery

## INTRODUCTION

Joinery is mostly straightforward when comprised of rectangular shapes (windows, doors, etc) or rhomboidal shapes (basic stairs and balustrades, etc) and a general knowledge of plain geometry is usually sufficient. However, maintenance- and refurbishment-work of traditionally-built period properties, public buildings and churches, etc, can involve replacing joinery items with curved components – such as a door- and doorframe-head with a Gothic- or Tudor-arch shape, etc – (covered in Chapter 6) – and this demands a greater knowledge of setting out geometrical shapes.

## BASIC ARCH-GEOMETRY DEFINITIONS

*Figure 8.1*: This shows the basic definitions (explained below) related to the setting out of a segmental-headed doorframe.

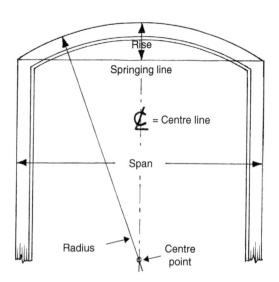

Figure 8.1 Basic arch-geometry definitions.

- *Springing line*: a horizontal reference or datum line at the base of an arched shape (from which the shape *springs*).
- *Span*: the overall width of an arched shape across the springing line.
- *Centre line*: a vertical setting out line equal to half the span.
- *Rise*: a measurement on the centre line between the springing line and the highest point of the arched joinery-shape.
- *Intrados*: the underside of the arched joinery-shape.
- *Extrados*: the topside of the arched joinery-shape.
- *Crown*: the highest point on the extrados.
- *Centre* or *centre point*: the pivoting or compass point of the radius.
- *Radius*: the geometrical distance of the centre point from the concaved shape.

## BASIC SETTING OUT TECHNIQUES

Before proceeding, a few setting out techniques in geometry must be understood.

### Bisecting a line

*Figure 8.2(a)*: This means dividing a line, or distance between two points, equally into two parts by another line intersecting it at right-angles. Figure 8.2(a) illustrates the method used. Line AB has been bisected. Using A as centre set a compass to any distance greater than half AB. Strike arcs $AC^1$ and $AD^1$. Now using B as centre and the same compass setting, strike arcs $BC^2$ and $BD^2$. The arcs shown as broken lines are only used to clarify the method of bisection and need not normally be shown. Now draw a line through the intersecting arcs $C^1 C^2$ to $D^1 D^2$. This will cut AB at E into two equal parts. Angles $C^1$ EA, $BEC^2$, $AED^1$ and $D^2$ EB will also be 90° angles; this also being, therefore, a useful practical way to create large right-angles.

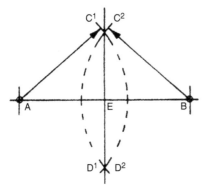

Figure 8.2 (a) Bisecting a line.

## Bisecting an angle

*Figure 8.2(b)*: This means cutting or dividing the angle equally into two angles. Figure 8.2(b) shows CAB as the angle to be bisected. With A as centre and any radius less than AC or AB, strike arc DE. With D and E as centres and a radius greater than half DE, strike intersecting arcs at F. Join AF to divide the angle CAB into equal parts CAF and FAB.

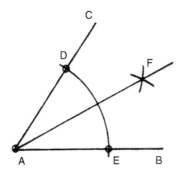

Figure 8.2 (b) Bisecting an angle.

## SEMICIRCULAR ARCHED SHAPE

*Figure 8.3*: Span AB is bisected to give C on the springing line. With C as centre, the semicircle is described from A to B.

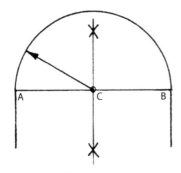

Figure 8.3 Semicircular arched shape.

## SEGMENTAL ARCHED SHAPE

*Figure 8.4*: Span AB is bisected to give C on the springing line. The rise at D on the centre line can be at any distance from C, but must be less than half the span. Bisect the 'imaginary' line AD (not normally shown when you become geometry-literate) to intersect with the centre line to establish E. With E as centre, describe the segment from A, through D to B.

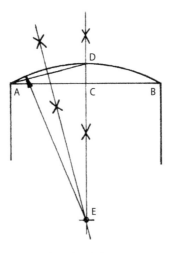

Figure 8.4 Segmental arched shape.

## DEFINITION OF ELLIPTICAL ARCHED SHAPES

*Figures 8.5(a)(b)*: An ellipse is the geometrical name given to the shape produced when a cone or cylinder is cut (theoretically or in reality) by a geometric plane (or an actual saw), making a smaller angle with the base than the side angle of the cone or cylinder. The exception is that when the cutting plane (or saw) is parallel to the base, true circles will be produced.

## AXES OF THE ELLIPSE

*Figure 8.5(c)*: An imaginary line (like a laser-beam or axle of a wheel) that passes through the exact cylindrical centre of a cone, cylinder or sphere, is known as an axis and the shape or space around (at right-angles to) the axis is equal in all directions. But when cones or cylinders are cut by an angled plane, as illustrated below in Figure 8.5(a) and (b), the shape or space on each side of the axis enlarges – but

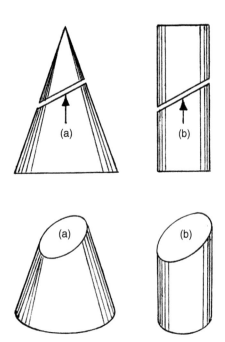

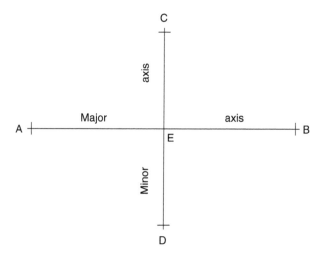

Figure 8.5 (a) Cone-produced ellipse and (b) cylinder-produced ellipse.

C

axis

Major                    axis

A                          B
         E

Minor

D

Figure 8.5 (c) Axes of the ellipse: ABCD = ellipse axes; ABC or ABD = semi-ellipse axes; AE or EB = semi-major axis; CE or ED = semi-minor axis.

only in one direction; and this forms an *elliptical* shape that now requires (for reference purposes) two axes – like vertical planes – at right-angles to each other, passing through the centre. The long and the short lines (planes) that intersect through the centre of ellipses are referred to as the *major axis* and the *minor axis*.

As illustrated in Figure 8.5(c), the axes on each side of the central intersection, by virtue of being halved, are called *semi-major* and *semi-minor* axes. The

semi-elliptical arched shape is so called because only half of the ellipse is used.

# TRUE SEMI-ELLIPTICAL ARCHED SHAPES

*True* semi-elliptical shapes, such as produced by 1) the *intersecting-lines method*, 2) the *intersecting-arcs method*, or 3) the *concentric-circles method*, are not shown here because they are not normally used by carpenters when producing wooden arch-centres for bricklayers. This is because these true methods of setting out do not give the bricklayer the necessary centre points as a means of radiating the geometric-normal of the voussoir-joints. And the arch-shaped frames, doors or windows, that invariably have to relate to brick-built arches, must follow the same setting out. However, before moving on to the more common *approximate* semi-elliptical shapes (which give the bricklayers their centre points), three practical methods of setting out *true* semi-elliptical shapes are shown below.

# SHORT-TRAMMEL METHOD

*Figure 8.5(d)*: Draw the major and semi-minor axes as illustrated and to the span and rise required. Obtain a thin lath or narrow strip of hardboard, etc as a trammel rod. Mark it as shown, with the semi-major axis $A^1 E^1$ and the semi-minor axis $C^1 E^2$. Rotate the trammel in a variety of positions similar to that shown, ensuring that marks $E^1$ and $E^2$ always touch the two axes; then mark off sufficient points at $A^1/C^1$ to plot the path of the semi-ellipse. In technical drawing (on a small scale) a flexi-curve aid or French curves can be used to link up the points and describe the semi-ellipse – but in a workshop (full-size scale) situation, I have used a long, narrow strip of hardboard as a flexible aid, with another person marking the

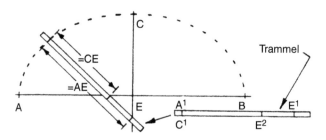

Figure 8.5 (d) True semi-ellipse by short-trammel method.

curve (or parts of it) as and when the hardboard aid is in position.

left and right, as indicated at HIJ, to describe a true semi-ellipse.

## LONG-TRAMMEL METHOD

*Figure 8.5(e)*: This is similar to the previous method, except that the semi-major and semi-minor axes form a continuous measurement on the trammel rod; the outer marks thereon are moved along the axes, while the inner mark, 0 (zero), plots the path of the semi-ellipse. This method is better than the previous one when the difference in length between the two axes is only slight. Note that the springing line on each side of the span – and the rise of the semi-minor axis – has to be extended to accommodate and relate to the trammel rod.

## APPROXIMATE SEMI-ELLIPTICAL ARCHED SHAPES

These approximate semi-elliptical shapes, as previously mentioned, are preferred for brick or stone arches, as they simplify the setting out and give the bricklayer or stonemason definite centre points from which to check the radiating geometric-normal lines of the voussoir joints.

## THREE-CENTRED METHOD

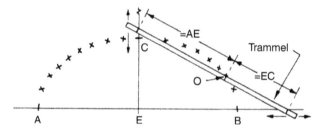

Figure 8.5 (e) True semi-ellipse by long trammel method.

## PIN-AND-STRING METHOD

*Figure 8.5(f)*: This method uses focal points on the major axis (the springing line). These are shown here as F and G – and either point equals AE or EB on a compass, struck from C to give F and G. To describe the arch shape, drive small, round-headed wire nails into points F, C and G, with their heads protruding. Pass a piece of string around the three nail-shanks and tie tightly. Make a simple pencil-jig, if possible and cut a notch in a pencil – as shown at $C^1$. Remove the nail at C, replace with pencil-and- jig and rotate to

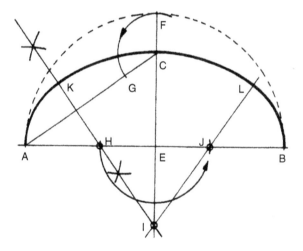

Figure 8.5 (g) Approximate semi-ellipse by three-centred method.

*Figure 8.5(g)*: Set out (as a *rod* on a white-emulsioned sheet or half-sheet of hardboard) the major axis, equal to the span and bisect this to create the semi-minor axis, equal to the rise required, as described initially. Draw a diagonal line from A to C (the chosen or given rise). With E as centre, describe semi-circle AB to create F. With C as centre, strike an arc from F to give G. Bisect AG to give centres H and I. With E as centre, transfer H to create J. Draw a line through IJ to give L. HIJ are the three centres required to describe the semi-ellipse. Draw sector-shapes AK from H, KL from I, and LB from J, to cut through the rise at C and complete the required shape.

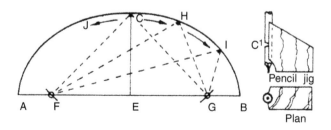

Figure 8.5 (f) True semi-ellipse by pin-and-string method.

# FIVE-CENTRED METHOD

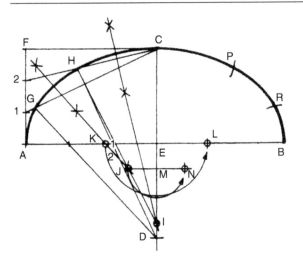

Figure 8.5 (h) Approximate semi-ellipse by five-centred method.

*Figure 8.5(h)*: Set out the major and minor axes as described above. Draw lines AF and CF, equal to CE and AE, respectively. Divide AF by three, to give A12F. Draw radials C1 and C2. With centre E and radius EC, strike an arc at D on the centre line. Divide AE by three to give A12E. Draw line D1 to strike C1 at G, and line D2 to strike C2 at H. Bisect HC and extend the bisecting line down to give I on the centre line. Draw a line from H to I. Now bisect GH and extend the bisecting line down to cut the springing line at K and line HI at J. IJK are the three centres to form half of the semi-ellipse. The other two centres are transferred as follows: with centre E, transfer K to give L on the springing line; draw a horizontal line from J to M and beyond; with centre M, strike an arc from J to give the centre N. To transfer the normal lines G and H, strike arc CP, equal to CH, then BR, equal to AG. To describe the semi-ellipse, draw sector-shapes AG from centre K, GH from centre J, HCP from centre I, PR from centre N, and RB from centre L.

# DEPRESSED SEMI-ELLIPTICAL ARCHED SHAPE

*Figure 8.5(i)*: To achieve this arched shape, a very shallow rise is used. The geometry is exactly the same

as that used for the *three-centred method* described above under that heading at 8.5(g).

# EQUILATERAL GOTHIC ARCHED SHAPE

*Figure 8.6(a)*: The radius of this arched shape, equal to the span, is struck from centres A and B to an apex point C. The extended line AD highlights a geometric *normal* to the curve – and a line at right angles to this, as shown, is known as a *tangent*. The normal lines E, F, G, H, I, J, K, L are indicated by broken lines to give an appreciation of the bricklayer's use of a centre point from which to line-up the joints of the brick-arch voussoirs. Incidentally, points A, B and C of this arch-shape, if joined by lines instead of curved arcs, form an equilateral triangle, where all three sides are equal in length and contain three 60° angles.

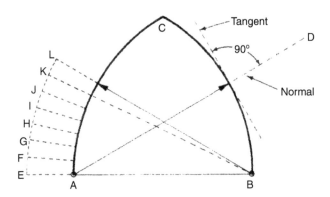

Figure 8.6 (a) Equilateral Gothic arched shape.

# DEPRESSED GOTHIC ARCHED SHAPE

*Figure 8.6(b)*: This *depressed* shape is sometimes referred to as an *obtuse-* or *drop-Gothic arch*. The centres for striking the shape come within the span, on the springing line. First, bisect the span AB to give the centre line through E. With a compass less than

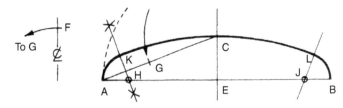

Figure 8.5 (i) Depressed, approximate semi-ellipse by three-centred method.

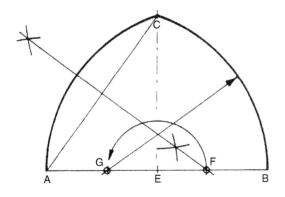

Figure 8.6 (b) Depressed Gothic arched shape.

AB, strike the rise at C from A. Alternatively, mark the chosen or given rise at C from E – which must not be less than AE or BE. Draw line AC and bisect to give centre F on the springing line. With centre E, transfer F to give centre G. Strike sector-curves AC from F and BC from G.

# LANCET GOTHIC ARCHED SHAPE

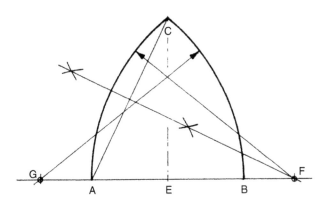

Figure 8.6 (c) Lancet Gothic arched shape.

*Figure 8.6(c)*: The centres for this arched shape are outside the span, on an extended springing line. First, bisect the span AB to give the centre line through E. With a compass set to more than AB, strike the rise at C from A. Alternatively, mark the chosen or given rise at C from E. Draw line AC and bisect to give centre F on the extended springing line. With E as centre, transfer F to give centre G. Strike sector-curves AC from F and BC from G. Note that the line AC need not actually be drawn, once you become geometry-literate.

# TUDOR ARCHED SHAPES

## Variable method

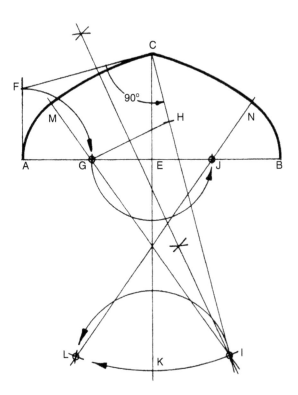

Figure 8.7 (a) Tudor arched shape – variable method.

*Figure 8.7(a)*: This method of setting out Tudor arched shapes is generally regarded as the best method and can be used to meet a variety of given or chosen rises. The geometry, which appears complex, is usually mastered when practised a few times.

First, draw the span AB and bisect it to give an extended centre line through and below E. Mark the chosen or given rise at C. Draw a vertical line up from A to give F, equal in height to two-thirds the rise (CE). Join F to C. At 90° to FC, draw a line down from C. With a compass setting equal to AF, and A as centre, transfer F to give G. With the same compass setting, mark H from C on line CI. Draw a line from G to H and bisect it; extend the bisecting line down until it intersects with line CI to give centre I. Draw a line from I, extended through G on the springing line. With E as centre, transfer G to give J on the springing line. Again with E as centre, transfer I, through K, to strike an arc at L. With K as centre, transfer I to give centre L. Draw a line from L to extend through J on the springing line. To form the arched shape, strike sector-curves AM from G, MC from I, CN from L and NB from J.

## Fixed method

*Figure 8.7(b)*: This method is simpler and can be used when the rise is not given, or is not critical and the only known information is the span. First, draw span AB and divide by four to give DEF. Draw vertical lines down from D and F. With D as centre, transfer F to intersect the vertical line to give G as a centre. With F as centre, transfer D to intersect the other vertical line to give H as a centre. Draw diagonal lines from H and G, extending through D and F on the springing line. To form the arched shape, strike sector-curves AI from D, IC from H, BJ from F and JC from G. Note that because the four centres form a square below the springing line, this setting out is also referred to as a *box-method Tudor shape*.

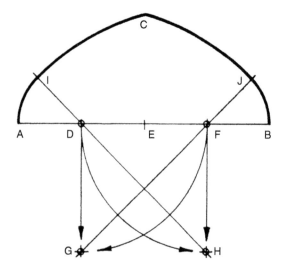

Figure 8.7 (b) Tudor arched shape – fixed or box method.

## Depressed Tudor arched shape

*Figure 8.7(c)*: First draw span AB and divide the springing line by six to give DEFGH. Draw vertical lines down from E and G. With centre D, transfer H down to O, and with centre H, transfer D down to O. Draw diagonal normal lines through DO and HO, extending down to intersect the vertical lines at K and L, and extending up past the springing line to establish I and J. To form the arched shape, strike sector-curves AI from D, IC from L, BJ from H and JC from K. Note that division of the span can be varied to achieve a different arched shape, as can the angles of the normal lines at D and H, drawn here at 60°. For example, 75° would make the arched shape more depressed.

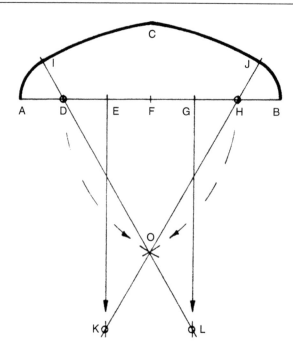

Figure 8.7 (c) Depressed Tudor arched shape.

## Straight-top Tudor arched shape

*Figure 8.7(d)*: First draw span AB and divide by 9. Mark one-ninth of the span from A to give D, and one-ninth from B to give E. With a protractor, or a modern roofing-square with a degree facility (if setting out full-size on a joinery rod), set up and mark diagonal normal lines passing through D and E at 78° to the springing line. With centre D, strike sector-curve AF, and with centre E, strike curve BG. From F and G, draw straight crown lines at 90° to the two normal lines, to intersect at apex C.

Note that the position of the centres D and E can be varied to achieve a different visual effect, as can the angles of the normal lines at D and E, drawn here

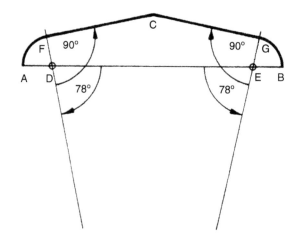

Figure 8.7 (d) Straight-top Tudor arched shape.

at 78°. However, the straight-top crown lines must always be tangential (at 90°) to the normal lines.

## SERPENTINE OR EYEBROW ARCHED SHAPE

*Figure 8.8*: This shape is sometimes encountered in joinery and cabinetmaking design, usually as the lower edge of top door-rails and – of course – the top edge of the door-panels. First draw span AB, equal to the reduced door-width between the top rail's shoulder lines. Establish centre line DG and mark the chosen or given rise at C from E. Mark F from A on the springing line, equal to about one-sixth of the span. Draw diagonal line CF (if you need actual lines when bisecting) and bisect it to cut the centre line at G. Draw horizontal line HI, equal to the rail's depth at AH and BI. Draw lines GH and GI (shown here as broken lines for clarity). These important lines are geometric normal lines to the curves required in reverse positions (*cyma reversa*) and each curve must stop at these lines to reverse the radiating direction. HIG are the three centres. To form the serpentine shape, strike sector-curves AJ from H, JC from G and BK from I. Note that this setting out is flexible and – providing the cyma reversa principle is not violated – a variety of low- or high-rise serpentine/eyebrow shapes can be achieved.

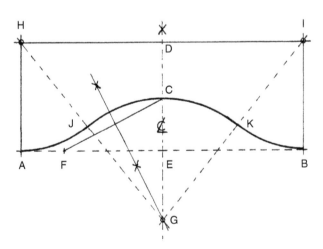

Figure 8.8 Serpentine or Eyebrow arched shape.

## TEMPLATES AND GEOMETRIC-NORMAL CENTRE-FINDERS

*Figures 8.9(a)(b)(c)*: Mindful of the fact that in maintenance- and refurbishment-work, replacement,

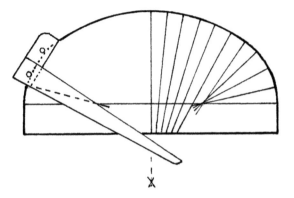

Figure 8.9 (a) Semi-elliptical shaped hardboard template showing the site-marked springing line; the purpose-made geometric centre-finder in position for further marking; and a number of centre-finding lines marked from the 'finder' on the right of the centre line, showing two centres found.

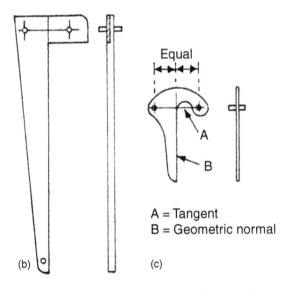

A = Tangent
B = Geometric normal

Figure 8.9 (b) Front- and end-elevational views of a purpose-made geometric centre-finder (of, say, 900mm length), joined by a glued-and-screwed half-lap joint and with tangentially positioned, projecting wooden dowels of 9mm diameter; and (c) Similar views of the small, plywood or steel centre-finder.

shaped window frame- and doorframe-heads usually have to be fitted to the underside of existing arches, where the original setting out method is unknown, the tradesperson usually has to resort to making a hardboard template on site. By trial-and-error marking, cutting and 'easing' (planing), the template is fitted as close as possible to the underside of the arch, but back in the workshop, it will be necessary to *improve* the curved shape to enable more precise working templates or jigs to be made. The *improving* is best done by taxing your knowledge of arch geometry and using it in reverse order to reproduce the original shape – as explained below.

For example, if you were making a shaped-headed frame from an imprecise hardboard template of an obvious-looking semi-elliptical arch shape, you must first assume it to be a three-centre arch and set about finding the three centres. This will enable you to check (and improve the curvature very slightly) of the template-shape by pivoting over the extremities of the template with a beam compass or radius rod from the three located centres. Figure 8.9(a) shows the hardboard template displaying *centre-finding* lines marked on it by a purpose-made wooden square which I call a *geometric centre-finder*. A more detailed illustration of this is shown in Figure 8.9(b). In wood-turning, a purpose-made centre-finder, made of wood or steel – identical in principle to mine, but much smaller – is used to find the centres of circular items to be 'turned' on the lathe. It is simply called a *centre-finder* and is shown at 8.9(c).

## PRACTICAL COMPASSES

*Figure 8.10*: To set out full-size curved shapes for jigs and templates, etc, a beam compass or a radius rod is required. As illustrated, the beam compass consists of a pair of *trammel heads* or *beam-compass heads* and a length of timber to be used as the beam; say of 40 × 20mm prepared hardwood section. To improvise, a *radius rod* can be easily made, consisting of a timber lath (of, say 40 × 10mm prepared section), with a panel pin or a small round wire-nail through one end and the other end drilled to hold a pencil firmly. Of course, the main disadvantage of a radius rod is the omission of a fine-adjusting facility (such as the eccentric point on a trammel head) and having to alter the pin or nail's position for every different setting.

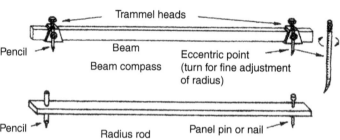

Figure 8.10 Details of a beam compass and an improvised radius rod.

# 9

# Traditional saw sharpening of non-hardpoint saws

## INTRODUCTION

Hardpoint, throwaway handsaws have virtually super-seded traditional handsaws for the following reasons: 1) they are cheap to buy; 2) they retain their original sharpness for a long time if used without hitting metal objects such as hidden nails, and 3) when the saw is blunt or damaged, there is no need to lose time in sharpening or paying for the saw to be sharpened. However, traditional non-hardpoint saws (that require to be sharpened when blunt) are still being sold. I know a number of craftsmen – and therefore suspect that there are many others – who still retain tradi-tional tenon-, panel-, crosscut- and rip-saws in their toolkits or workshops and look upon them as their 'good saws', even though they mostly use hardpoint saws for convenience!

This chapter, therefore, is for those hand-skills' enthusiasts who want to attempt and master the high degree of skill and judgement required in successful sharpening of traditional saws.

## SEQUENCE OF OPERATIONS

There are four separate operations involved for saws in a bad condition; these are known as 1) topping, 2) shaping, 3) setting, and 4) sharpening – and, for best results, they are performed in that sequence. However, if a saw is in a good condition, has not been neglected or abused, but has lost its edge through normal use, then the action needed is less drastic and it will only require sharpening.

## TOPPING

*Figures 9.1(a)(b)(c)*: Normally, the points of the saw's teeth conform to a straight line or – with some saws – a slightly segmental curve in the length of the saw. If, however, through lack of the joiner's time or skill, the

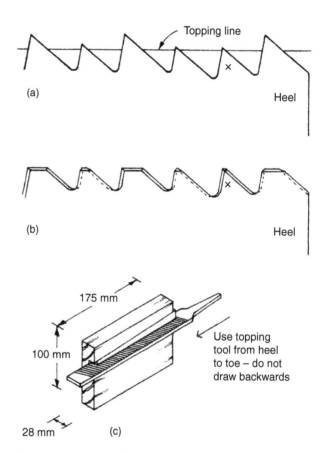

Figure 9.1 (a) Misshapen teeth highlighted by an imagi-nary 'topping line'; (b) the teeth after topping to the lowest tooth at 'x'; and (c) The topping tool.

saw is sharpened roughly on a number of occasions, the line or camber will lose its original good shape and the teeth will become misshapen and unequal in size and height. Such teeth are known as *dog teeth*. To remedy these faults, the teeth must be reshaped and the first step is known as topping. This means running a flat mill file over the points of the teeth until the tips of the lowest teeth have been 'topped' by the file, as indicated in Figures 9.1(a) and (b), and the overall shape has been regained in the length of the saw. To achieve this, periodic sightings – looking along the line of teeth at eye-level – must be made during the filing operation.

## Topping tool

*Figure 9.1(c)*: To assist the filing operation, a wooden block grooved to take the mill file and a wooden wedge, should be made. The complete assembly, illustrated in Figure 9.1(c), is known as a *topping tool*. Its main advantage lies in keeping the file – by virtue of the block being pressed against the saw blade – at right angles to the saw. Finally, it must be borne in mind that excessive topping creates extra work in the next operation, *shaping*.

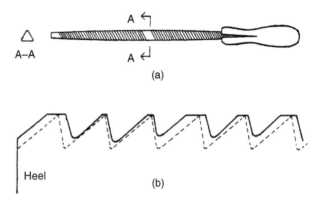

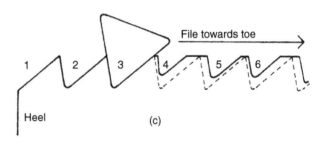

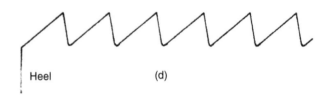

# SHAPING

*Figure 9.2(a)*: The operation of *shaping* (and *sharpening*) is carried out with a saw file. Such files are equilaterally triangular (60°), slim-tapered (as illustrated) in various lengths, single- or double-ended and are fitted into plastic or wooden saw-file handles. 150mm double-ended files are recommended for dovetail-, tenon- and panel-saws and 200mm or 225mm double-ended files for crosscut- and rip-saws.

## Shaping technique

*Figures 9.2(b)(c)*: When shaping, the filing action is always square across the saw blade and follows in every consecutive gullet from the heel on the left to the toe on the right. The idea is to eliminate the dog teeth and create evenly shaped teeth leaning towards the toe of the saw at the correct pitch and points-per-25mm. Pitch refers to the angle-of-lean given to the front cutting edges of the teeth (as shown at the end of this chapter). The recommended pitch angle for rip saws is 87°, for crosscut saws is 80°, and for panel-, tenon- and dovetail-saws is 75°. Although these angles are not critical to a few degrees, until experience is gained in angle-judgement, the required angle can be set up on an improvised template or sliding bevel to test the degree of accuracy in initial shaping.

# FILING ACTION

*Figures 9.2(c)(d)*: When filing, the handle should be held firmly in one hand and the end of the file steadied and weighted with the thumb and first two fingers of the other hand. All the strokes should be forward-acting and *not* drawn backwards. Take great care in establishing the correct angles and shapes on the first few teeth and then familiarize yourself with the feel of the file resting in a corrected gullet. Maintain this feel (and go back occasionally to regain it, if

**Figure 9.2 (a)** Double-ended saw file; **(b)** Broken lines indicating the shaping-outline to be judged; **(c)** Recommended saw position for shaping sequence; and **(d)** the final, shaped appearance.

necessary) as you continue the shaping operation – and combine it with close, visual appraisal of the shape and pitch of the teeth in relation to the shiny, flat areas on the tips. These flat tips are produced in the topping operation and will gradually diminish as the gullets deepen, indicating that the shaping must stop immediately the shiny tips are removed. Finally, it helps to rub chalk on the file occasionally to reduce the tendency of the file to become clogged with metal particles.

# SETTING

*Figure 9.3*: This next operation, known as *setting*, refers to the bending of the upper tips of the teeth, every other one, out from the face of the saw on one side and then setting the alternate row of teeth on the other side. This is done with a pliers-type tool known as a *saw-set*. The idea is that the cut – or kerf – made by the saw is slightly wider than the thickness of the saw blade, to create clearance and facilitate an easy sawing action. However, too much *set* can be a disadvantage, as the saw tends to run adrift in an oversize saw kerf. For this reason, it is advisable to set the saw slightly less than the numbered setting indicated on the saw-set and only reset the saw when it is really necessary – not every time the saw is re-sharpened.

## Saw-set tool

*Figures 9.3(a)(b)*: As illustrated, the saw-set has a knurled hand-screw controlling a wheel-shaped, bevelled anvil – the edge of which is numbered with different settings, relative to points-per-25mm – pp25 or ppi (points-per-inch) – and not *teeth* per 25mm (which is always one less than pp25 and if used, therefore, creates more set). The anvil has to be adjusted so that the required pp25 numeral (6 pp25 setting shown in Figure 3(b)'s plan view) is exactly opposite the small plunger which ejects when the levered handle is squeezed.

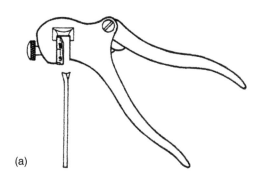

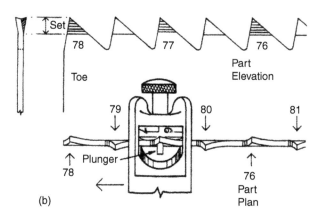

Figure 9.3 **(a)** The saw-set positioned over the saw's teeth; and **(b)** Part end- and front-elevation of the alternate, numbered settings and part plan-view of the saw-set in its 77th setting-position.

## Setting technique

To set the saw, hold it under your arm, with the handle in front, the teeth uppermost, and place the saw-set on the first tooth facing away from the plunger. Looking down closely from above, squeeze the saw-set's levered handle firmly and then carefully repeat this operation on every other tooth thereafter from the heel towards yourself and the saw's toe at your rear. Then turn the saw around so that the handle is now at the rear, under your arm, and set the alternate row of teeth from the toe towards yourself and the heel – which is gradually worked out from the under-arm position. The saw is now ready for sharpening. Note that when setting old saws, squeeze the levered handle very gently, as the metal becomes brittle with age and the teeth can snap off in the setting operation.

# SHARPENING

*Figure 9.4*: This final operation is concerned with creating sharp-cutting front edges and sharp points to the outer tips of the teeth, by filing every other gullet at an angle to the face of the saw on one side – and then the alternate row of gullets at an opposing angle to the face of the saw on the other side, as indicated in the part-plan and elevation in Figure 9.4(b).

## Sharpening theory

*Figures 9.4(a)(b)*: The method of sharpening shown in Figure 9.4(b) is for saws designed to cut across the grain, such as crosscut-, panel-, tenon- and dovetail-saws, the theory being that the sharp-pointed outer tips of the inverted vee-shaped teeth act as knives cutting two close lines across the timber. Short-grained pieces of fibre between the lines break up as the saw moves forward; the broken fibres are collected in the gullets as *sawdust* and released when the saw passes through the timber. Rip saws are similarly sharpened on alternate sides, but the sharpening angle is square or almost square across the saw, as indicated in Figure 9.4(a). This eliminates the pointed outer tips required for crosscutting and produces square-tipped teeth which provide a scraping/shearing action necessary for effective rip-sawing along the grain.

## Sharpening technique

*Figure 9.4(c)*: When sharpening, take care not to lose the basic shape of the teeth; this is best achieved

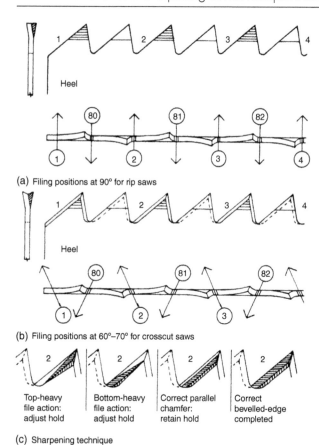

(a) Filing positions at 90° for rip saws

(b) Filing positions at 60°–70° for crosscut saws

Top-heavy
file action:
adjust hold

Bottom-heavy
file action:
adjust hold

Correct parallel
chamfer:
retain hold

Correct
bevelled-edge
completed

(c)  Sharpening technique

**Figure 9.4 (a)** Part-plan and elevation of filing positions at 90° for rip saws; **(b)** filing positions at 60° to 70° for crosscut saws (and others such as panel-, tenon- and dovetail-saws); **(c)** part-elevation of a sharpening technique referred to below.

by gaining the feel of the file and by keeping the back edges of the teeth constantly in view. The first stroke of the saw file, at an angle of 60° to 70° to the saw-face, should show a parallel chamfer on the back edge of the tooth. If not parallel, then adjust the file accordingly on subsequent filing-strokes and stop immediately the chamfer-edged tooth becomes completely bevel-edged – as illustrated step-by-step in Figure 9.4(c).

This normally takes from two to four strokes and, once established, each gullet should receive the same number of strokes thereafter. This promotes a rhythmic filing action necessary for speed and accuracy. If a saw is only being re-sharpened, whereby the edges of the teeth are already bevelled, it helps to top the saw *very lightly* with the topping tool. The idea is to maintain a good line of tipped-teeth by aiming to split the shiny flat spots in half when sharpening alternately from one side of the saw and then remove the remaining halves when sharpening from the other side.

## Saw-sharpening frame

To enable the above procedures to be carried out, different kinds of hand-made saw-sharpening frames have been devised over many years. And although they seem to be known under different names such as *saw-stocks*, *saw-chops*, *saw-vices* or *saw horses* (the latter being a dual reference to *saw-stools*) in different localities, they all share a common function in aiming to act as a vice in gripping the *entire length* of the thin saw-blades just below the gullets. However, the wooden jaws of these vices are quite commonly referred to as *saw-chops* in themselves – and the universally popular way of holding them together on each side of the saw-blade seems to be to wedge them into vee-shaped slots in a simple wooden frame. Although the elevation- and plan-views of the saw-chops and upper legs of the saw-sharpening frame are shown in the following illustrations detailing the *sharpening procedure*, precise details of design and construction of a saw-sharpening frame are given at the end of this chapter.

## Sharpening procedure (for right-handed persons)

*Figure 9.5(a)(b)*: Cramp the saw high in the saw-chops – as illustrated at (a) – and top it lightly with the topping tool. Next, reposition the saw so that only about 4mm remains between the top of the saw-chops and the base of the gullets, as at (b). With the saw handle to your left, rest the sharpening-frame against a bench, etc, with good light in front and above the saw.

## Starting position

*Figures 9.5(c)(d)*: Take up your position against the sharpening-frame by resting your right foot on the bottom rail (the *foot rail*), with your knee pressing against the top rail (the *knee rail*). This is to keep the sharpening-frame steady during the sideways-thrusts of the filing action. Start to file at the heel (near the saw-handle), as illustrated at (d), in the gullet affecting the back of the first tooth leaning away – noting that the file, at an angle of about 60° to 70° to the saw face, *always points towards the saw handle*. After two or three forward strokes, aimed at '*splitting the shiner*', repeat the action in every other gullet thereafter, moving rhythmically towards the toe of the saw on your right.

## Changing position

*Figure 9.5(e)*: When the last quarter of the saw's length is reached, it will be found easier to switch the leg position and support the sharpening-frame with

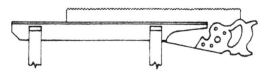

(a) Topping position

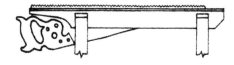

(b) Sharpening position

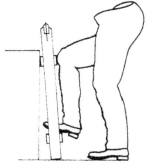

(c) Starting position

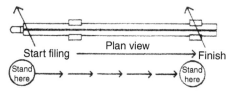

(d) Filing

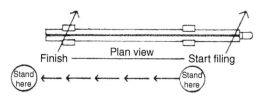

(e) Change position

Figures 9.5 (a)(b)(c)(d)(e) Topping and sharpening positions.

the left leg's foot and knee. When completed, turn the frame around so that the saw handle is now to your right, as illustrated at (e). Once more, support the frame with your left leg's foot and knee and start to file at the heel again (near the handle), in the gullet affecting the back of the first tooth leaning away, remembering that the file *always points towards the saw handle*. With the same number of file strokes per gullet used on the first operation, aim to remove the remaining half of the 'shiner' and produce sharp points whilst moving rhythmically towards the toe of the saw on your left, as indicated at (e). Again, in the last

quarter, switch the leg position and support the saw stocks with the right leg's foot and knee.

## Alternative sharpening procedures

If preferred, the saw-sharpening frame can be held rigid and upright by cramping one of the legs in a bench vice (and by packing-out the other, unsupported leg against the side of the bench – to eliminate shuddering). Or, alternatively, without any saw-sharpening frame, the jaws of the saw-chops alone can be held in a bench vice. But, unless you are a very short person, this last set-up is back-aching and not conducive to a relaxed, rhythmic filing action. Traditionally, to give the saw-chops more height above the bench vice, they consisted of a T-shaped, hinged arrangement resembling a crocodile document-clip. The stem of the T was adjusted for height in the bench vice.

## Saw-sharpening frame construction

*Figures 9.6*: The material used is usually softwood and can be of prepared or sawn finish. The wooden

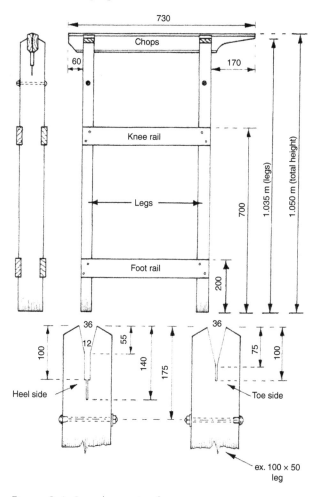

Figure 9.6 Saw-sharpening frame.

jaws (*saw-chops*), which hold the saw in the frame, are made from ex 75mm × 25mm material (preferably of *redwood* or *hardwood*). The frame is made up of two ex 100mm × 50mm *legs*, four ex 75mm × 25mm *cross rails*, two on each side, acting as *foot-* and *knee-rails*, optional *top rails* above the knee rails and an optional *diagonal brace*.

## Joinery-shop or site-made sharpening frames

A diagonal brace was usually used on site-made sharpening frames to strengthen the simple construction of the surface-nailed rails. On workshop-made frames, the rails were usually housed and screwed into the legs and so there was no need for a brace. Also, the saw-chops were often made of hardwood and a coach bolt was inserted through the edge of each leg, near the top as illustrated, to offset the tendency for the legs to split when the saw-chops were driven in to the vee-cuts to hold the saw.

## Preparing the legs

*Figure 9.6*: The logical first step in making a joinery-shop sharpening frame is to prepare the legs. Because it is important that a person takes up the correct posture at the sharpening-frame, the height is critical and ideally should be to suit the individual. The total height of 1.050m given here would be suitable for a tall person of about 1.830m (6ft).

## Provision for saw-chops

*Figure 9.6*: The vee-cuts made in the legs to receive the saw-chops should be marked out in relation to the vee-shaped housings to be cut in the sides of the saw-chops. As illustrated in Figure 9.6, the vee-shape should promote a slow (gradual) wedge action, as opposed to a less acute angle that would not take such a good grip on wedge shapes driven into it. At the base of the vee-cuts, a saw cut is made to the depths shown, to house the upturned blade of the saws being sharpened. One leg, on the side chosen to take the heel of the saws, must have a further slot of about 12mm width to accommodate back-saws such as tenon saws.

## Alternative rail fixings

*Figure 9.7*: Next, the rails are cut to length and, as illustrated, can be surface-fixed, part-housed or

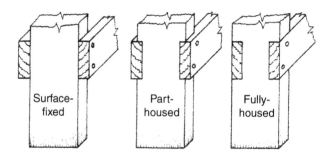

Figure 9.7 Alternative rail fixings to legs.

fully-housed into the legs and fixed with 65mm round-head wire nails or minimum 45mm × 10 gauge countersunk screws. Their position, to act as foot- and knee-rails to keep the sharpening-frame pinned against the bench, is an important ergonomics feature which enables the operative's hands to be free for the sharpening operation – thus avoiding the awkward uprightness of being held in a bench vice.

## Tapered housings in saw-chops

*Figure 9.8*: After fixing the rails, the saw-chops can be prepared to fit the legs and must be of sufficient length to accommodate the longest saw. The tapered, vee-shaped housings are marked and cut to fit the vee-cuts already established and cut in the legs (as detailed in Figure 9.6). As detailed, these housings must be of unequal distances from the ends to allow for a greater projection of the saw-chops from the leg that was slotted to take the back-saws.

## Allowance for saw handles

As illustrated in Figures 9.6 and 9.8, the handle-end of the projecting saw-chops are cut to a shape that will house the handles of the various types of saws that might require sharpening. This shape can vary, or be varied, according to the range of saws that need attention.

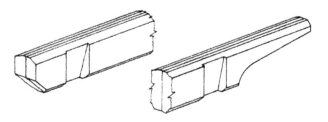

Figure 9.8 Tapered housings and end-shaping to saw-chops.

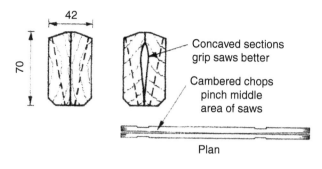

**Figure 9.9** Optional refinements to saw-chops.

## Optional saw-chops' details

*Figure 9.9*: As illustrated, the sectional shape of the chops varied between a site-made and a joinery-shop made sharpening-frame. Obviously, the former would have needed to be simple and the latter could be more refined. A concave shape, as shown in the second cross-sectional illustration, helped to pinch the saw just below the gullets of the teeth, thereby eliminating distracting movement of the blade during sharpening. Also, a further refinement of the saw-chops was achieved if each inner face was planed very slightly *round* in length, to a convex, or cambered shape (as illustrated in the *plan* view). When such chops were tightened into the vee-shaped slots, the cambered shapes compressed the middle area of the blade that lacked the grip of the leg slots.

## Recommended leg-bolts

Finally, to reinforce the legs from the likelihood of splitting, it is recommended that two 9mm diameter coach bolts should be inserted in the legs, as shown in Figure 9.6. After insertion and tightening of the nuts onto washers, any surplus bolt should be cut off

with a hacksaw and filed or hammered to remove any dangerous burrs or sharp metal edges.

# HANDSAWS

The following illustrations and information regarding traditional handsaws refers to the pitch and the number of teeth for each type of saw, but it also refers to the recommended sawing-angles – which are still relevant to all hand-saw users, regardless of the saws being of a traditional- or hardpoint-type.

## Crosscut saw

*Figure 9.10*: As the name implies, this is for cutting timber across the grain. Blade lengths and points-per-25mm (pp25) or ppi (points-per-inch) vary, but 660mm (26 inches) length and 7 or 8 pp25 are recommended. All handsaw teeth on traditional-type saws contain 60° angular shapes leaning, by varying degrees, towards the toe of the saw. The angle-of-lean relative to the front cutting edge of the saw is called the *pitch*. When sharpening saws, it helps to know the required pitch. For crosscut saws the pitch should be 80°. When crosscutting, the saw (as illustrated) should be at an approximate angle of 45° to the timber.

## Panel saw

*Figure 9.11*: This saw is for fine crosscutting and is particularly useful for cutting sheet material such as plywood or hardboard. It has a blade length of 560mm (22 inches), 10 pp25 and a 75° pitch is recommended. When cutting thin manufactured boards (plywood, hardboard, MDF, etc) the saw should be used at a low angle of about 15 to 25°.

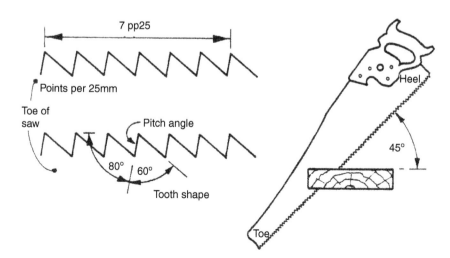

**Figure 9.10** Crosscut saw.

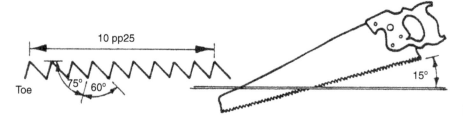

Figure 9.11 Panel saw.

## Tenon Saw

*Figure 9.12*: Because of its brass or steel back, this saw is sometimes referred to as a *back saw*. It is technically a general purpose bench saw for fine cutting. The brass-back type, as well as keeping the thin blade rigid, adds additional weight to the saw for easier use. The two most popular blade lengths (professionally) are 300mm (12 inches) and 350mm (14 inches). The 250mm (10 inches) saw is less efficient because of its short stroke. On different makes of saw, the teeth size varies between 13 and 15 pp25. For resharpening purposes – although dependant upon your skill and eyesight – 13 pp25 is recommended, with a pitch of 75°.

Figure 9.12 Tenon saw.

## Rip saw

*Figure 9.13*: This saw is used for cutting along, down, or with the grain – and is no doubt the least used nowadays because of the common use of machinery.

However, it is very useful in the absence of electrical- or battery-power – or if you need some exercise. It has a blade length of 660mm (26 inches) 5 or 6 pp25 and a pitch of 87° is recommended. When ripping along the grain, the saw should be used at a very steep angle of about 60 to 70° to the timber. Because of the square-edged teeth and pitch angle, this saw cannot be used for crosscutting.

Traditional saws should be kept dry if possible and lightly oiled, but if rusting does occur, soak liberally with oil and rub well with fine emery cloth.

# HARDPOINT HANDSAWS AND TENON SAWS

*Figures 9.14(a)(b)(c)(d)(e)*: These modern throwaway saws have high-frequency hardened tooth-points which stay sharper for at least five times longer than conventional saw teeth. Three shapes of tooth exist; the first, referred to as universal, conforms to the conventional 60° tooth-shape and 75° pitch; the second, known as the *fleam* tooth, resembling a *flame* in shape (hence its name), with a conventional front- pitch of 75°, an unconventional back-pitch of 80°, giving the fleam-tooth shape of 25°; the third, referred to as *triple-ground*, has razor-sharp, circular-saw-type tooth geometry, enabling a cutting action on both the

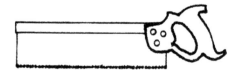

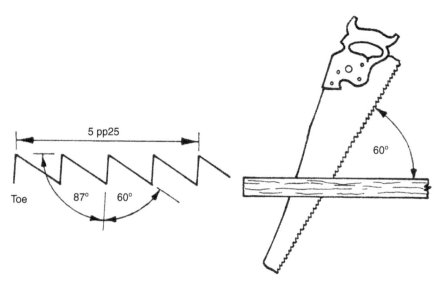

Figure 9.13 Rip saw.

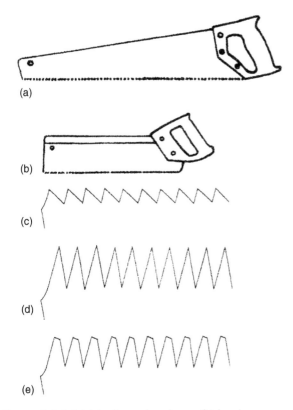

Figure 9.14 (a) Hardpoint handsaw; (b) hardpoint tenon saw; (c) universal tooth-shape; (d) fleam tooth-shape; (e) triple-ground tooth-shape.

push and the pull strokes. Most of the handsaws are claimed to give a superior cutting performance across and along the grain. Some saws in the range have a Teflon-like, friction-reducing coating on the blade to eliminate binding and produce a faster cut with less effort.

## Range of sizes

These saws usually have plastic handles – some with an improved grip – and a 45° and 90° facility for marking mitres or right angles. The handsaw sizes available are 610mm × 8 pp25, 560mm × 8 pp25, 508mm × 8 pp25, 560mm × 10 pp25, 508mm × 10 pp25, 480mm × 10pp25, 455mm × 10 pp25 and 405mm × 10 pp25. Tenon saw sizes available are 300mm × 13 pp25, 250mm × 13pp25, 300mm × 15 pp25 and finally 250mm × 15 pp25. The three recommended saws from this range would be the 610mm × 8 pp25 and the 560mm × 10 pp25

black-coated handsaws and a 300mm × 13 pp25 tenon saw.

## Pullsaws

These lightweight, unconventional saws of oriental origin, cut on the pull-stroke, which eliminates buckling. They can be used for ripping or crosscutting. The unconventional precision-cut teeth, with three cutting edges, are claimed to cut up to five times faster, leaving a smooth finish without breakout or splintering. The sprung-steel blade is ultra-hardened to give up to ten times longer life and can easily be replaced at the push of a button. Replacement blades cost about two-thirds the cost of the complete saw, but a complete saw is relatively inexpensive.

# GENERAL SAW AND FINE SAW

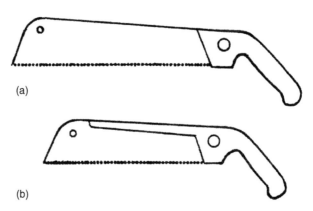

Figure 9.15 (a) General carpentry saw; (b) fine-cut saw.

*Figure 9.15*: Only two saws from the range are covered here. The first is called a *general carpentry saw* and has a 455mm blade × 8 pp25. This model comes in two other sizes, 380mm × 10 pp25 and 300mm × 14 pp25. The latter is recommended for cutting worktops and laminates without chipping. The second model is called a *fine-cut saw* and has a half-length back or full-length back support – and is said to surpass conventional tenon saws. This model comes in two variations, one with a fine-cut blade of 270mm × 15 pp25, the other with an ultra-fine blade of 270mm × 17 pp25.

# Index